CW01455584

# Private Militaries and the Security Industry in Civil Wars

# Private Militaries and the Security Industry in Civil Wars

## Competition and Market Accountability

SEDEN AKCINAROGLU AND
ELIZABETH RADZISZEWSKI

**OXFORD**
UNIVERSITY PRESS

# OXFORD
### UNIVERSITY PRESS

Oxford University Press is a department of the University of Oxford. It furthers
the University's objective of excellence in research, scholarship, and education
by publishing worldwide. Oxford is a registered trade mark of Oxford University
Press in the UK and certain other countries.

Published in the United States of America by Oxford University Press
198 Madison Avenue, New York, NY 10016, United States of America.

© Oxford University Press 2020

All rights reserved. No part of this publication may be reproduced, stored in
a retrieval system, or transmitted, in any form or by any means, without the
prior permission in writing of Oxford University Press, or as expressly permitted
by law, by license, or under terms agreed with the appropriate reproduction
rights organization. Inquiries concerning reproduction outside the scope of the
above should be sent to the Rights Department, Oxford University Press, at the
address above.

You must not circulate this work in any other form
and you must impose this same condition on any acquirer.

Library of Congress Cataloging-in-Publication Data
Names: Akcinaroglu, Seden, 1972– author. | Radziszewski, Elizabeth, author.
Title: Private militaries and the security industry in civil wars: competition and
market accountability [Seden Akcinaroglu, Elizabeth Radziszewski].
Description: New York : Oxford University Press, [2020] |
Includes bibliographical references and index.
Identifiers: LCCN 2020014157 (print) | LCCN 2020014158 (ebook) |
ISBN 9780197520802 (hardback) | ISBN 9780197520826 (epub)
Subjects: LCSH: Private military companies. | Private security services. |
Mercenary troops. | Civil War.
Classification: LCC UB149 .A43 2019 (print) |
LCC UB149 (ebook) | DDC 355.3/4—dc23
LC record available at https://lccn.loc.gov/2020014157
LC ebook record available at https://lccn.loc.gov/2020014158

1 3 5 7 9 8 6 4 2

Printed by Integrated Books International, United States of America

*Elizabeth dedicates this book to her family—*
*Jolanta, Zbigniew, Adam, Ela, Kaylen, and James—*
*with gratitude and love*

*Seden dedicates this book to her parents whose unconditional love and support*
*have made it possible to write this book. You made me believe in today when*
*I thought I would never get there*

# Contents

Contents

# Tables and Figures

## Tables

## Figures

# Acknowledgments

We are indebted to a number of individuals who made this project possible. In particular, we thank former employees of private military and security companies, whose anonymity we seek to preserve, for sharing their candid views of the industry. We are grateful to Leticia Armendariz for discussing NGOs' role in monitoring, and to Doug Brooks for valuable ideas. Christopher Mayer, Director of Armed Contingency Contractor Policies and Program, provided important insights on contractor accountability. His patience and openness have greatly enriched our research and writing process. We thank Ulrich Petersohn and Adam McMahon for help during various stages of writing and research and are incredibly grateful to our editor at Oxford, Angela Chnapko, for believing in the project. Research and writing were generously supported by a fellowship from Rider University.

We are most thankful to our families for lifting us up and reminding us of what truly matters. Elizabeth thanks Jamie Loper for his support and God for inspiration.

Princeton, NJ, 2020

# Abbreviations

| | |
|---|---|
| AFRICOM | The United States Africa Command |
| ANSI | American National Standards Institute |
| BAPSC | British Association of Private Security Companies |
| COIN | Counterinsurgency |
| COW | Correlates of War |
| CRG | Control Risks Group |
| DoD | Department of Defense |
| DSL | Defense Systems Limited |
| EO | Executive Outcomes |
| FMLN | Frente Farabundo Martí para la Liberación Nacional |
| GSG | Ghurka Security Guards |
| ICoC | International Code of Conduct for Private Security Service Providers |
| ICoCA | International Code of Conduct Association |
| ISO | International Organization for Standardization |
| ISOA | International Stability Operations Association |
| KBR | Kellogg Brown and Root |
| KMS | Keenie Meenie Services |
| MPRI | Military Professional Resources Inc. |
| NGO | Non-Governmental Organization |
| PAE | Pacific Architects and Engineers |
| PKK | Kurdistan Workers' Party |
| PMC | Private Military Company |
| PMF | Private Military Firm |
| PMSC | Private Military and Security Companies |
| PRIO | Peace Research Institute Oslo |
| PROGO | Project on Government Oversight |
| PSC | Private Security Company |
| PSCAI | Private Security Company Association of Iraq |
| ROC | Reconstruction and Coordination Center |
| RUF | Revolutionary United Front |
| SCEG | Security in Complex Environments Group |
| SIGIR | Special Inspector General for Iraq Reconstruction |
| UCDP | Uppsala Conflict Data Program |
| UK | United Kingdom |
| UN | United Nations |
| US | United States |

# 1

# External Interventions and Civil Wars: Why Focus on Non-State Actors?

In 2010, General David Petraeus, then top US commander in Afghanistan, remarked that contracting security services "represents both an opportunity and a danger."[1] Whether they train police forces in Afghanistan and Iraq, provide logistical support to African governments hoping to gain an upper hand over rebel combatants, or prepare government forces to manage terrorist threats, private contractors are present in numerous conflict zones around the world. Often well trained and organized, private military industry's employees are some of the leading security experts in the world. Unlike mercenaries fighting individually or as part of loosely organized groups, today's private military and security companies (PMSCs) are corporate entities that deliver military and security services for monetary compensation, maintain permanent locations, and have a hierarchical corporate structure. Some are listed on stock exchanges.

With American soldiers stretched thin worldwide and state failure posing security threats that the international community has been, at times, unwilling to address, the demand for private contractors to enter some of the most complex and dangerous intrastate conflicts has risen to unprecedented levels. Sensing new and lucrative opportunities, private military and security providers began to flood the market. Since the end of the Cold War in 1993, a time when the United States downsized its military and major powers withdrew from weak states, leaving behind a vacuum of power, the market for private contractors has rapidly grown, as has the companies' ability to provide increasingly more sophisticated, innovative, and diverse services. Corporate actors are ubiquitous; they help governments train their military, deliver intelligence, plan for combat, transport personnel, perform risk assessment, and they have occasionally engaged in direct combat when fighting alongside the government's forces.[2] The industry's expansion and reports of human rights abuses in Iraq and Afghanistan have spurred the need to evaluate the industry's impact on conflicts. Are these contractors effective in

*Private Militaries and the Security Industry in Civil Wars.* Seden Akcinaroglu and Elizabeth Radziszewski, Oxford University Press (2020). © Oxford University Press. DOI: 10.1093/oso/9780197520802.001.0001.

bringing an end to violence or has the desire for profit and limited accountability tainted their usefulness? What are the conditions under which PMSCs sometimes help with conflict termination, while in other cases their impact is limited or negative?

This book's major focus is on understanding the conditions under which private military companies are more likely to be effective in terminating civil wars, the dominant type of armed conflict in the world since the end of World War II. Civil wars continue to persist—in 1990 there were 49 such conflicts with a at least 25 battle-related deaths while in 2014 there were 42.[3] The potential for state failure that could create conditions conducive to fighting is unlikely to abate in the near future. Such state failure is most often associated with the presence of weak political institutions that are characteristic of anocracies or hybrid, semi-authoritarian regimes.[4] Marred by corruption, these regimes usually lack the capacity to maintain social order. Global trends in governance show that the number of hybrid regimes has dramatically increased after the end of the Cold War, and while there have been periods of some decline, the growth has been positive since 2007.[5] In response to rebel attacks, many governments have turned to private contractors for help. But reliance on PMSCs has not been restricted merely to weak states. The United States, Great Britain, and other major powers have employed private contractors to fulfill their foreign policy interests such as securing gains against insurgents in Iraq and Afghanistan. In light of such developments, understanding intrastate violence and instability increasingly requires that we understand the role of external, non-state actors in shaping war dynamics.

Much of what we know today about external interventions into civil wars concerns the role of states and international organizations in molding conflict dynamics. The general argument, supported by empirical findings, is that external interventions prolong fighting because they either invite counter interventions or give the combatants a chance to regroup, rearm, and continue their armed struggle.[6] More recently, there has been a growing interest in the causes and consequences of interventions by non-state actors such as foreign volunteer fighters that have been present in Syria's civil war and diaspora's military support and lobbying efforts aimed at international actors with the power to influence civil war dynamics.[7] Aside from such actors' presence in conflict zones, there is also the rise in the number of interventions by PMSCs. In 2006 nearly 300 PMSCs intervened in civil wars around the globe, an unprecedented increase since 1984, when only one company delivered services in a conflict zone. While scholars interested in the private military industry have debated the positive and negative aspects associated with the proliferation of private contractors, systematic global analysis of conditions under which PMSCs have contributed to the cessation of violence in civil wars or alternatively prolonged it remains elusive. To date, there have been only a few efforts[8] to study PMSCs' impact on civil war dynamics despite the fact that civil wars remain the most frequent form of armed conflict, and PMSCs intervene in such conflicts at a much greater rate than states in the post–Cold War era. This book, therefore, is an opportunity to bridge the gap between these two areas of research and encourage further exploration of the linkage between private military and security industry's presence in civil wars and conflict termination.

To understand PMSCs' impact on the termination of civil wars and the preservation of peace in the conflicts where they intervene, we need to identify conditions under which some PMSCs are more militarily effective in contributing to the government's victory in a civil war from conditions under which the effectiveness is limited. Because PMSCs are profit-oriented businesses, the temptation to be less effective exists, especially if such behavior can pass undetected or be justified to the client.[9] Scholars interested in private military and security industry recognize that stronger monitoring mechanisms are vital to better regulate the industry's behavior in conflict zones.[10] Furthermore, while some researchers recognize the positive contribution of PMSCs to global security,[11] many view the industry with suspicion, seeing it as a threat to state authority and stability.[12] This project departs from the assumption that interventions by profit-driven military and security corporations are likely to destabilize the security environment.

First, while the cases of PMSCs' behaving badly make the news and ignite critical debates, PMSC intervention on behalf of the state is sometimes associated with shorter wars. Our data indicate that PMSCs do, in fact, contribute to conflict termination under certain conditions even when we control for alternative explanations. PMSCs have different corporate structures and some experience more competitive pressure than others in specific circumstances. These factors will exert varying impact on PMSCs' behavior and levels of accountability in conflict zones, meaning that under some conditions PMSCs can bring positive contributions to conflicts in which they operate.

Second, we argue that the competitive market pressure that PMSCs face creates a monitoring mechanism and that variation in two dimensions of market pressure, the industry-wide competition and the competitive environment specific to a given conflict, helps explain why some companies are more conscious of competitors in ways that push them to be more effective in securing gains for the government and hence accountable to the client. When companies are accountable to the client they are more likely to display higher level of military effectiveness, which enables them to play a more significant role in shifting the balance of power in favor of the state and thus contribute to swifter termination of conflicts. In other words, the market serves as a regulatory mechanism that deserves much more attention than currently given in existing scholarly and policymaking debates.

By focusing on market pressure, the competitive environment in which PMSCs operate, we can predict when companies are more likely to be effective and hence exert positive influence on conflict dynamics. PMSCs' behavior is not always capricious. Consider, for example, the company's corporate structure, which may evolve and change depending on market pressure. Merely looking at a company's ownership structure, whether it is publicly or privately held, tells us a lot about its level of transparency and accountability and hence its self-imposed regulations about behavior in conflict zones. By uncovering the company's structure we can better understand what kind of clients the company will work for and how it will fulfill its contractual agreement. All of these factors, in turn, will enable us to predict whether the company can offer positive contribution to the termination of civil wars.

## Market Pressure and PMSCs' Accountability in Civil Wars

Previous work on PMSCs' behavior in conflict zones posits that because the industry is not responsive to the state in the way that public armies are, it suffers from severe accountability deficit. In the absence of accountability, PMSCs roam the conflict zones as free agents, operating according to their own rules and norms. Since PMSCs are in the business of making profit, these rules are likely to reflect their own interests more so than those of their clients. The discrepancy between the interests of the client and those of the company could undermine rather than improve the security in conflicts. Yet this argument rarely considers the possibility that the market environment in which PMSCs operate can curtail their willingness to maximize profits using any means. Peter Singer (2003),[13] for example, argues that PMSCs should be studied in a context with either similar military companies or with overarching business models rather than analyzed in isolation. What this implies is that companies are not always free to behave in a way that best suits their short-term interest. PMSCs are aware of the growing competition and the extent to which their performance in one conflict zone could impact business opportunities elsewhere. A Defense Systems Limited (DSL) employee once remarked that "when we sneeze in Africa, we catch a cold in Asia."[14] Companies' reputations matter in securing future contracts and thriving in an increasingly crowded industry.[15] Consequently, market pressure can, at times, make it too costly for some companies to risk their long-term survival for the sake of greater and more immediate profit that could nevertheless be short term.

In the past, limited emphasis on norm emergence in the private military and security industry and the relative newness of industry meant that fewer companies were concerned with reputational consequences of their behavior. Sometimes in this environment, it was easier for companies to neglect strategic, long-term planning for winning the war in favor of short-term gains in the battle that often come with such practices as cutting costs by relying on poor-quality equipment, engaging in fraudulent practices, or displaying limited concern for humanitarian laws—all the factors that we will argue in Chapter 2 can negatively affect military effectiveness in civil wars. As the industry began to expand and the negligent behavior of some providers became more widely covered in the media, PMSCs became more sensitive to reputational costs. This is especially true for companies established along corporate lines with the goal of long-term survival. Although companies may still survive despite their connection to human rights abuses or fraud, doing so takes serious rebranding that often leads to modification of behavioral practices in tandem with greater reputational concerns. Such was the case, for example, with Erik Prince's Xe Services that was formed in light of Blackwater's tarnished reputation. Under new CEO Ted Wright, Xe, which changed to become Academi in 2010, has replaced over 80% of its former instructors to avoid indoctrination into belligerent military culture as part of its rebranding stage. To improve its image, Wright established a code of conduct that all employees are required to follow.[16]

Even in the informal market structures, those outside of the US and the UK, where business is done with limited domestic regulation, international companies with corporate structures must still demonstrate effectiveness or risk either being replaced by a competitor or failing to secure a contract with another client that might be concerned with its lackluster performance. When the market's limitation is visible it is in the context of the so-called fly-by-night companies that are temporary in existence.[17] These companies, such as France's Secrets, are not designed with a long-term corporate development in mind[18] and thus may exhibit little or no interest in upholding reputational concerns. Such companies may thus exploit the client by delivering poor-quality services without considering how such practices would impact their ability to secure a contract renewal. Overall, we assume that most companies organized along corporate lines exhibit some minimal concern for long-term reputational costs in light of growing market pressure while recognizing that temporary companies would be somewhat less susceptible to these constraining mechanisms.

We further build on the work of Spear (2006)[19] and Avant (2005),[20] who argue that consumer demand and determination to preserve their competitive advantage may serve as a mechanism for regulating PMSCs' effectiveness in conflict zones. The market imposes a degree of discipline on the companies and pushes them to strive for reliability, effectiveness in military missions, and even greater respect for human rights as a means of protecting their reputation so vital for securing new business. At the same time, Spear (2006)[21] notes that market pressure is also imperfect and may not restrain profit-hungry companies from modifying self-regulation to fit their interests. Like Spear (2006), we posit that competitive pressure is much more dynamic and multilayered than often assumed. For example, consumers of security services may not always be in a position to detect the extent to which their clients are performing at a limited level of effectiveness, and in such instances competitive pressure may exert lower regulatory impact. Given that most civil wars occur in weak states whose governments lack the capacity to wage an effective struggle against the rebels, their ability to monitor PMSCs' performance will be challenging. A PMSC could still help the government win the war but because the government lacks the monitoring capacity, it will not perceive the extent to which the company could have helped terminate the conflict sooner. Cockayne (2007) acknowledges such a possibility by recognizing that PMSCs can "co-opt their principals, by developing excessive 'epistemic power' over their principals' preferences."[22] Expanding on previous works, we push the argument further and explore conditions under which market pressure could increase PMSCs' accountability in conflict zones when clients lack strong monitoring capacity. Put simply, the degree to which self-regulation works is highly conditioned on the clients' ability to hold companies accountable for their actions—the latter being linked to effective monitoring—and the different reputational consequences associated with limited accountability. Therefore our contribution is to develop a causal story that improves our understanding of when market pressure is more likely to make a difference in changing PMSCs' incentives from opportunism to military effectiveness while concurrently highlighting the market's limitations in such endeavors.

Our focus in this study is on any international PMSC that operates as a legally registered international corporate entity with a clear business structure and that delivers military and security services for monetary compensation.[23] Although military and security companies in our data set range from those that engage in combat and provide military and security training, to firms responsible for logistical tasks, communications, and intelligence,[24] we examine the impact of any type of company because even logistical assistance and other non-direct-combat-related tasks can help a warring party obtain advantage over an enemy and affect war dynamics. Weak military effectiveness in areas outside of direct combat involvement could have an adverse impact on conflict termination. Carmola (2012)[25] argues that revelations of the use of torture to interrogate detainees in Iraq by private contractors hired for intelligence gathering might have contributed negatively to US counter-insurgency efforts in that conflict, a war in which human rights abuses by foreigners have alienated the locals and pushed them into the hands of the insurgents. While companies with different service provision are included in our main analysis because they have the potential to shape civil war termination, nevertheless we are still able to examine whether differences in the type of services that PMSCs deliver could explain variation in their contribution to conflict termination (Chapter 5).

We explore two dimensions of market pressure that increase effectiveness among security providers in ways that could push them to be more accountable to the client. In Chapter 2 we identify two elements of military effectiveness—material/individual capabilities and corporate professionalism—that we argue are associated with PMSCs' ability to help governments fulfill the missions in conflict zones. While material and individual capabilities are more obvious dimensions of military effectiveness and have received more extensive coverage in the literature, corporate professionalism is not frequently linked to such effectiveness. For us, corporate professionalism espouses a commitment to the rule of law that limits fraudulent behavior and promotes adherence with international humanitarian laws. There has been a significant debate about the extent to which adherence with international humanitarian laws, in particular, matters as a component of military effectiveness and how best to balance it with other, somewhat more aggressive strategies.[26] In light of the evidence we discuss in Chapter 2, we advocate for a balanced approach that suggests flexibility yet one which makes human rights protection an important goal in securing the support of local population in the fight against the rebels. Overall, we argue that variation in PMSCs' commitment to corporate professionalism will be shaped by divergent market pressure that companies face.

First, we consider the specific conflict environment in which PMSCs operate. While most scholars agree that companies operating in a neoliberal market face growing competition,[27] we argue that competitive pressure is, in fact, much more dynamic than commonly assumed. There is the overall trend of having more companies around at the industry level, which puts pressure on everyone to stand out when bidding for contract. But there is also the pressure to distinguish oneself in the midst of the competitive environment that companies face when fighting in a given conflict zone alongside other security providers. Some clients may choose to hire more than one provider, and this dramatically alters the environment in which PMSCs function. As the competitive field becomes crowded, the client not only has the opportunity to compare every company's effectiveness more directly but, more important, can also, as we will argue in Chapter 3, rely on companies to informally monitor each other's behavior in conflict zones. The competitive pressure for contract renewal and preserving reputation is likely to drive this informal, monitoring phenomenon and push companies to demonstrate a high level of military effectiveness, which entails a lower likelihood of subverting the client's interests through the use of low-quality equipment, engagement in fraud, and indiscriminate killings of civilians. Like Cusumano (2009),[28] we argue that companies can engage in self-regulation. But our concept of self-regulation is much more aggressive than joining industry-wide trade organizations to signal commitment with good practices, many of which, including compliance with international humanitarian laws, can be linked to better military effectiveness in conflicts and thus greater accountability to the client. When joining an industry-wide trade organization, a PMSC signals some level of commitment to these practices, but failure to be part of such an organization is less likely to impose significant costs on the company. By contrast, improving their effectiveness in a conflict zone when faced with a crowded local market may be a necessity for many PMSCs because failure to demonstrate effectiveness could be exploited by competing companies, yielding both immediate and long-term costs. Put simply, competitive pressure is not merely limited to the bidding phase but continues even after companies intervene in a conflict zone.

We then consider a second dimension of market accountability, the industry-wide or global competition that has affected decisions about PMSCs' corporate structure. We argue that in light of growing competition at the industry level, some companies have altered their business models drastically to gain advantage over their rivals. Existing debates on the private military industry rarely analyze the impact of corporate structure on PMSCs' behavior, even though McIntyre and Weiss (2007)[29] acknowledge that variation in corporate tradition among PMSCs is vast with potential to impact such activities as the companies' ability to better screen their employees. In other words, it could matter whether you hire ArmorGroup, a company that was listed on the London Stock Exchange until 2008 before it was acquired by a publicly traded security giant G4S, or Stabilico, a firm that McIntyre and Weiss (2007)[30] argue is merely a group of mercenaries who label themselves as a "company." Built into this argument is the assumption that companies with more-established corporate traditions are likely to exhibit greater accountability because they may invest in practices that could improve their effectiveness in conflicts.

We posit that embracing a more transparent corporate structure by becoming a publicly traded company, while mostly motivated by a desire to secure capital, can credibly communicate to potential clients the companies' level of accountability and serve as a way for a PMSC to distinguish itself from other competitors in an increasingly crowded marketplace of security providers. Unlike private companies, publicly held businesses sell part of their ownership to the public and are thus listed on the stock market. By contrast, private companies are usually owned by a small group of initial founders and private shareholders.

Such divergent corporate organizations create varying structures of accountability with a profound impact on the companies' business practices. Not only are publicly traded companies required to disclose their financial earnings to the public, but because the public has a vaster interest in such companies due to their ownership rights, the interest to investigate and monitor the companies could be greater among the press. Shooting at civilians or wasting the client's valuable resources on "ghost" employees carries a risk for any PMSC, but it is likely to be greater for those that are accountable not only to the clients but also to the public. The public, after all, is the investor and investor image constitutes a critical component of corporate reputation.[31]

Criminal behavior, when uncovered, could diminish public trust in the company, contribute to the decline in its stock value, and trigger a massive loss of capital. This type of behavior signals that the company is less likely to be effective in securing gains for the client. Whether the media focuses on the company's effectiveness in conflicts or solely on the ethics of behavior in light of more frequent coverage of human rights abuses in general in the post–Cold War era,[32] bad press generates more vulnerabilities for publicly traded PMSCs than other types of companies. Put simply, it would be more difficult, though not impossible, for companies to recover from scandals and secure future contracts especially at a time when the industry boasts multiple and capable players who are eager to take on new clients.

The book delves deeper into the mechanism of accountability that stems from corporate ownership structure. Chapter 4 addresses, for example, the reason why a company would take the risk of selling part of its ownership to the public and expose itself to public audiences in unprecedented ways. In what ways does this exposure affect corporate professionalism? Do the data on human rights abuses and fraudulent practices committed by PMSCs in conflict zones show that publicly traded companies behave differently than privately held companies? This part of the book examines whether accountability-generating market forces lead to greater improvements in the area of corporate professionalism for publicly traded PMSCs than for private ones. We then explore how the difference in companies' corporate structure impacts their military effectiveness in terminating conflicts.

## Theoretical Significance, Overview of the Data, and Policy Implications

Given the proliferation of PMSCs in civil wars, understanding the conditions under which PMSCs' interventions can reduce the duration of wars and pave the road for long-term stability is clearly important for scholars interested in conflict management and resolution. By focusing on interventions by non-state actors other than international organizations, foreign volunteer fighters, and diaspora, the book offers new theoretical contributions to research on external interventions into intrastate wars. Our argument and empirical analysis enable us to examine how interventions involving PMSCs affect war termination and how they compare to the impact of interventions without PMSCs.

There is also a contribution to research on strategic third-party interventions, an area that is gaining more attention.[33] Akcinaroglu and Radziszewski (2005),[34] for example, argue that some states deliberately intervene to prolong civil wars in order to destabilize the government. Domestic instability, they argue, gives the intervening state leverage over its interstate rival, thereby creating an incentive to arm the rebels in ways that could jeopardize settlement. Similarly, PMSCs may have a strategic interest not to push for conflict termination immediately if doing so allows them to maximize profit. While it is in the PMSCs' interest to help the governments win, states are often not in a position to judge the time frame that is needed to secure victory. As such, private contractors may exploit the governments' inability to monitor the situation on the ground to deliver a "slow" victory. The book extends the findings from existing literature on intervention by investigating the effect of corporate actors' behavior on war dynamics, and conditions under which market pressure could deter opportunistic action by altering actors' strategies in ways that align more with the interests of their clients.

Second, the book's argument about the impact of competitive pressure on PMSCs' behavior in conflict zones has theoretical implications for those interested in delving more into the role of competition in shaping industry behavior. We relax the assumption that PMSCs strive for insecurity because it allows them to expand their profits. We argue that as business ventures PMSCs strive to secure profit, but that opportunity structures affect their behavior. Thus to understand the impact of these entities on conflict dynamics, it is imperative to place them in the context of the environment, the market that determines decisions about accountability. Far from being uniform, competitive pressure is experienced at different levels for different companies.

We distinguish, for example, between what we refer to as local, more immediate competition at the conflict level from the general competition that companies experience because of the industry's global expansion. With the exception of some closed markets, such as the one in China until recently or those that are dominated by criminal gangs that provide security and leave no opportunities for other players to freely enter the market (as in Guatemala), companies today face the burden of expanding global competition. The occurrence of this phenomenon has meant that a majority of companies have tried to show their uniqueness to secure contracts by diversifying their services, promising to use high-quality equipment, or claiming to attach value to international humanitarian laws. Yet monitoring problems have persisted.

Initiatives such as the Montreux Document (2008)[35] that outlines states' legal responsibility related to the use of PMSCs in armed conflict and the International Code of Conduct for Private Security Service Providers,[36] which has focused on increasing third-party monitoring of companies' actual capabilities and initiatives to uphold humanitarian laws, have not necessarily made it easier to monitor PMSCs' behavior in conflict zones. The more recent establishment of ANSI (American National Standards Institute)/ASIS International's 2012 Management System for Quality of Private Security Company Operations,[37] updated in 2017, and the International Organization for Standardization's (ISO) 2015 Management System for Private Security Operations[38] asks companies to develop criteria for improving compliance with international human rights laws but has a limitation when it comes to conflict zone monitoring, as there is no guarantee that companies will uphold the standards once they receive the contract.

We argue that global competition has had the biggest impact on those companies that have responded to the growing market pressure by becoming publicly traded. Because of their unique corporate status, the issues we delineated earlier become less problematic. By contrast, competition at the local level can exert greater pressure on a wider spectrum of companies, whether they are publicly traded or not, as long as those companies are operating alongside other firms in a given conflict. A company may be aware that new companies are popping up on the market and promise to send highly qualified men to a conflict zone to secure a contract during the bidding phase—in doing so it appears to be responding to *global or industry-level competition*. However, if it is a sole provider of services to a given government during an armed conflict it only needs to worry about delivering some, albeit not a high level of effectiveness because there is no other provider that they could be compared to and monitoring their performance in the field could be challenging. As such, the company may try to cut costs by sending less-qualified men on a mission who may very well end up helping the government make some progress. With a more creative and capable force, however, the said company might have been in a position to secure the gains faster and thus exhibit greater effectiveness in terminating the war. In this scenario the *local, conflict-level competition* is absent and there is little pressure on a company to maximize its effectiveness. Unless the company is publicly traded, in the absence of local competition there are limited assurances that a PMSC will be accountable to the client even when it has responded to global competition by, for example, purchasing better-quality weapons. A company may simply forgo the use of expensive equipment if it requires hiring more-qualified men to operate it or using the equipment for one client but not another.

Our argument about different dynamics of market pressure takes us into an area that is rarely explored by scholars interested in PMSCs. Understanding how companies respond to global vs. local competition allows us to gain more insights about companies' behavior in conflicts and what to make of PMSCs' attempts to respond to international initiatives aimed at better regulating the industry. To date, for example, there is no systematic analysis of how variation in PMSCs' corporate structure, in large part driven by global competition, affects companies' effectiveness in civil wars. Scholars acknowledge that corporate structure matters[39] yet widespread ambiguity surrounding the connection between business models and PMSCs' behavior in conflicts persists. By investigating a critical dimension of corporate structure, the company's ownership, we show that companies' business models exert a significant impact on the level of commitment toward practices that increase military effectiveness with positive consequences for the termination of violence and greater accountability to the client.

Third, in our attempt to push the research on PMSCs in a new direction, we also concentrate on systematically testing our arguments in ways that would enable us to make generalizations across time and space. Most of what we know about private military industry is based on case study research and policy recommendations for curbing private warriors' potential for opportunistic behavior.[40] This work is vast and valuable in helping us to better understand the opportunities and dangers that industry growth presents today. At the same, however, scholars have only recently begun to systematically analyze conditions that affect PMSCs' performance in ways that would allow us to deliver specific policy suggestions to future clients. For example, Petersohn (2017)[41] has examined how different types of PMSC intervention affect conflict intensity, while Tkach (2019)[42] examines how variation in contract specifications for PMSCs and intra-service competition impact levels of violence in Iraq. Still, much remains unknown about conditions under which PMSCs' presence can improve security on the ground. This book adds to this nascent body of research on PMSCs' impact on conflict outcomes by presenting novel data on PMSCs' involvement in all civil wars[43] around the globe from 1990 to 2008.[44] While notoriously difficult to collect because of secrecy, denials, and covert activities, our data address these concerns through systematic efforts to increase reliability using multilayered collection process and source verification. For example, in addition to relying on secondary sources, such as newspapers and reports, we occasionally consulted with PMSCs' employees to verify information about the timing of companies' intervention and the activities they were involved in. These data build upon data collection undertaken by the British Foreign and Commonwealth Offices for the years 1990–1999 in the African context,[45] Akcinaroglu and Radziszewski's (2013) data on Africa that cover 1984–2008,[46] the Private Security Database from Data on Armed Conflict and Security that features data on PMSCs' interventions into failed or failing states,[47] Tkach's (2019) governorate-level data on PMSCs in Iraq from 2004 to 2008,[48] and most recent events-based data addition by Avant and Neu (2019) that provides information on PMSC's activities in Latin America, Africa, and South East Asia from 1990 to 2012.[49]

Our data on all civil wars from 1990 to 2008 provide information on each company that has intervened in every conflict year, on the type of service that each intervening company has provided, and on the companies' corporate structure in both major wars with at least 1,000 battle deaths and minor wars with at least 25 battle deaths. When considering data sets that have expanded their focus beyond a single country or continent, our data have notable differences. For example, they differ from the Private Security Database by expanding its focus on PMSCs' interventions into all major and minor civil wars rather than concentrating only on the subset of cases involving failed or failing states. As such, we are able to provide insights on a more comprehensive set of cases of armed conflicts involving PMSCs' interventions on behalf of the government. Furthermore, new data we present in this book differ from Avant and Neu's (2019) data set in several areas. While Avant and Neu focus on a longer time span, our data are global in nature, covering all countries, with limitation for Iraq and Afghanistan.[50] Second, our data concentrate on international PMSCs' interventions into major and minor civil wars, while Avant and Neu code any activity of commercial security providers without a specific focus on different types of wars. Their Private Security Events Database also includes state and any non-state clients of any type of commercial security contractors, while ours concentrates on interventions on behalf of the government and rebels by a specific type of PMSC, international ones. Our data's focus is on international PMSCs, as we expect that such companies are likely to experience different reputational concerns from other types of commercial security providers and thus exhibit unique impact on conflict dynamics.

Finally, our data contribution extends beyond PMSCs' interventions and ventures into the area of companies' human rights abuses and fraud in the case of Iraq (2003–2019). While the data do not account for the duration of each PMSC's presence in Iraq, we note whether any PMSC that has operated in the country at any point has been connected to human rights abuses or fraud. These data allow us to examine the logic of our argument by empirically testing the mechanism through which we argue market pressure improves companies' performance. As we expect that competitive pressure positively impacts some companies' commitment to corporate professionalism, which we argue is an element of military effectiveness in civil wars, we are able to examine the validity of this point by focusing on whether publicly traded companies that alter their corporate structure in response to global competition are more committed to good practices than are private companies. Here, our data efforts on human rights abuses and fraud build upon Avant and Neu (2019), who code companies' connection to crimes but not in the context of Iraq.

Fourth, policy implications emerging from this study could help us to better manage private military and security contractors whose presence is unlikely to diminish in the future. Consider the case of US reliance on PMSCs. In 2009, the peak year for private security contractors' presence in Iraq, over 15,000 contractors were supporting the US Department of Defense's (DoD) war effort.[51] The numbers then went into decline, with 2,500 present in 2013. In 2016, Admiral Michael Rogers, commander of US Cyber Command, noted during a Senate hearing that private contractors made up 25% of his workforce.[52] The presence of private security contractors working for the DoD in Afghanistan peaked in 2012 at over 28,000. Although these numbers decreased substantially by 2016, they were again on the rise in 2017 and do not reflect contracts from the State Department and intelligence agencies. With President Donald Trump's announcement to limit US troop presence in Afghanistan in 2019,[53] it is likely that contractors will gain even greater presence in the country.

Beyond Iraq and Afghanistan, reliance on PMSCs in Africa has also been of growing interest to the US government. Contracts managed by the State Department and AFRICOM—US military command for Africa established in 2007—focus on contractors' assistance with military training and peacekeeping operations. Lastly, contractors have bolstered US efforts in Latin America through the training of Colombia's and Mexico's forces as part of antinarcotics and counterterrorism initiatives.[54] While the US government has awarded the greatest number of the most lucrative contracts to private military and security contractors, other governments across the globe have also sought the industry's services. In 2014, for example, Nigeria relied on private contractors to train the military in unconventional warfare against Boko Haram,[55] while the British government, the second-largest consumer of private contractors, spent nearly $300 million on contractors in Iraq between 2003 and 2008.[56]

Although recent years have witnessed the return of mercenary groups such as Russia's the Wagner Group in Syria's and Ukraine's civil wars and in parts of Africa,[57] the demand for services from international PMSCs, which are the focus of our book, is likely to continue in the future. Not only are such companies leading innovators in the military and security sphere,[58] they are also, overall, more likely to respect international humanitarian laws than mercenaries even if there is variation among international PMSCs' commitment to such laws. Both of these characteristics make them more desirable service providers to consumers that are interested in cutting-edge technology and new strategies, and to governments that are accountable to domestic audiences for respecting the rule of law. The latter has become especially relevant after the media's revelations of human rights scandals involving American PMSCs working for the US government in Iraq. A company interested in bidding for a contract with the US DoD must demonstrate certification with ANSI/ASIS or ISO Standards and show how it meets the requirements to assure human rights protection.[59] Mercenary groups do not have the structure to meet such certifications, and it is doubtful whether they would even seek it as they have different business goals than do international PMSCs in terms of long-term survivability. In light of these, it is not surprising that demand for international PMSCs is unlikely to diminish even as mercenaries have also returned to serve the needs of other clientele, usually politically weak states. According to Visiongain's (2019) report on the future of private military and security market growth, which focuses on international PMSCs such as Academi, CACI International, DynCorp, Erinys International, and Garda World Security Corporation, among others, the market is set to grow to $420 billion by 2029, with demand increasing at a high rate. The report indicates that established and emerging markets as well as commercial companies will seek assistance from private contractors due to layoffs of experienced military personnel and the need to manage the changing nature of armed conflict.[60]

Better management of international PMSCs is thus of interest to policymakers and scholars concerned with growing privatization of the state and foreign policy, phenomena that have led to non-state global actors taking over, to various extents, responsibilities previously embraced by states. There is considerable disagreement about the positive and negative consequences of the changes brought forth by globalization and the rise of non-state actors in terms of their impact on state sovereignty and respect for the rule of law.[61] Our book contributes to the broader theme pertaining to the privatization of the state by showing conditions under which a specific subset of non-state security providers, international PMSCs, could help terminate fighting that threatens to weaken the state domestically and internationally or alternatively prolong violent fighting among groups contending for state control.

The message of this study is that increased reliance on non-state actors can be a good or a bad thing depending on how the state manages its relations with such players. To this end, the book advocates greater awareness of market dynamics that PMSCs face to complement existing regulatory frameworks at the national and international level to increase the industry's accountability. Initial efforts in this area, especially early ideas about increasing accountability, have not always generated desirable results. Not only were initiatives such as reliance on embassy staff to monitor PMSCs costly to implement, but also they have not been strong enough to modify industry behavior in ways that could have significant impact on conflict dynamics. If anything, because of their leadership in military innovation and risk assessment, PMSCs have managed to influence national elites to grant them contracts to serve governments with poor human rights records. Such was the case when at one point MPRI (Military Professional Resources Inc.) lobbied the US government to allow the company to intervene in Equatorial Guinea despite laws that prohibit assistance to governments guilty of human rights abuses.[62] More recent attempts to certify PMSCs' adherence with humanitarian laws and establish appropriate norms of behavior—Montreux Document 2008, International Code of Conduct for Private Security Service Providers 2010, ANSI/ASIS Standards 2012/2017, and ISO Standards—while representing the most advanced approach to monitoring so far and being valuable in specifying what it means for the company to be accountable, face some notable challenges related to third-party monitoring of PMSCs in the field.

Our approach focuses on a complementary method to increase that sought-after accountability. Understanding the distinction between industry-wide competition and local competition that the companies encounter in a given conflict can help clients manage their contracts in ways that will maximize PMSCs' military effectiveness. The book argues that since market pressure exerts significant influence on PMSCs' behavior, potential clients should exploit ways in which they can take advantage of the competitive environment to strengthen PMSCs' effectiveness in conflict zones. This enables the government to regain control of its security goals rather than allow private military industry to hijack them for their own gain.

## Plan of the Book

In the next chapter we highlight the rise of the private military and security industry, focusing in particular on the arguments about the negative and positive aspects of the trend to outsource security to PMSCs. Unlike public forces, which are responsive to the state, employees of a private security firm are accountable to the company that hires them while the company itself is accountable to the client and its shareholders. These divergent mechanisms of accountability create more opportunities for private military and security industry to operate according to its own rules and interests, which sometimes deviate from the interests of the client. As such, this chapter investigates the potential for PMSC interventions to negatively affect the termination of violence. At the same time, this chapter also looks at the other side of the debate and highlights existing arguments about the positive impact of PMSC interventions. It focuses on ways in which private contractors can boost the government's power to offset rebel incursions and create an opening for the cessation of hostilities. In contrast to past studies, we advocate the need to move beyond the debates about the negative and positive aspects of PMSCs' interventions into conflicts and focus instead on uncovering conditions under which such interventions can have a divergent effect on conflict dynamics. Recent research is starting to move in this direction and we contribute to this emerging trend.

The key to improving PMSCs' contribution to conflict termination is to increase the actors' accountability to the client. In Chapter 2 we examine current approaches to doing just that and demonstrate how our approach complements existing efforts. We examine what it means for a PMSC to be militarily effective in civil wars by discussing the importance of capabilities and the previously overlooked dimension of military effectiveness: corporate professionalism. How is corporate professionalism useful in helping PMSCs secure gains against insurgents in ways that could help the governments terminate the war? In particular, in light of the growing debate about the benefits of population-centric vs. enemy-centric approaches to counter-insurgency, we present an extensive discussion on the connection between adherence with international humanitarian laws (one component of corporate professionalism) and military effectiveness in civil wars. The discussion then focuses on the second dimension of corporate professionalism: the connection between limited fraudulent practices and military effectiveness. We highlight how market pressure creates conditions for improving corporate professionalism, which in turn increases military effectiveness of PMSCs in ways that make them more accountable to the client. Lastly, we describe our data on PMSCs' interventions into civil wars in the context of major and minor wars. We present initial insights on some notable trends in PMSCs' interventions by looking at changes over time and the types of services that these actors have delivered in both types of conflicts.

Chapters 3 and 4 develop an argument about ways in which variation in two critical dimensions of market pressure affects PMSCs' performance. The main focus is on the monitoring mechanisms through which market pressure constraints PMSCs' incentives to perform less effectively in conflict zones. We emphasize two dimensions of market pressure: the local, competitive environment that companies encounter in a given conflict and global competition that has pushed some companies to shift their corporate structure, with implications for conflict performance. Our general argument is that not every PMSC will encounter the same level of market pressure, and this variation can be a good predictor of PMSCs' effectiveness in civil wars. Is it common for multiple companies to work in a conflict zone or do governments prefer to outsource their security needs to only one provider? How has the level of local competition changed over time? How common is it for companies to sell their ownership to the public and increase their accountability? Is this a rising trend or merely a series of isolated cases? We turn to our data to explore these questions.

One of the notable contributions of Chapter 4 comes from insights based on our new data on human rights abuses and fraud—a predictor of weak military effectiveness that stems from weak corporate professionalism—committed by PMSCs that have intervened in one prominent case, the war in Iraq. We are able to compare the difference in fraud and human rights abuses linked to publicly and privately held companies. These data enable us to specifically show how market pressure, competition at the industry level, has pushed some companies to adopt a more transparent corporate structure that is associated with higher levels of accountability in a case where data on human rights abuses are highly reliable, in large part due to multiple organizations' efforts to document these abuses. It is this accountability, we posit, that can make a difference in whether a PMSC that intervenes in civil wars exerts a positive or negative impact on conflict duration.

Chapter 5 presents empirical findings on the two dimensions of market forces and PMSCs' impact on the duration of civil wars. The analysis compares the impact of PMSCs on conflict duration in an environment where only one competitor operates to an environment where the number of providers increases. We also disaggregate the competition variable by looking not only at competition in a given conflict year based on the total number of intervening PMSCs regardless of the service they provide, but also at a competition among companies who provide the same type of service in a given conflict year. This enables us to examine if competition in one type of service is more important than competition in another. Our analysis also compares the impact of PMSCs' interventions into civil wars on the duration of conflicts with interventions by state actors and international organizations. Existing research finds that overall interventions prolong wars, but we demonstrate that interventions by PMSCs have a unique effect on conflict termination in comparison to other types of interventions. When local competition among security providers increases, PMSCs are more militarily effective, and their interventions have a positive impact on conflict termination in a specific subset of conflicts, major wars with over 1,000 battle deaths.

Our focus then moves to PMSCs' corporate structure and its impact on civil war duration. In the analysis, we investigate whether publicly held companies' presence in conflict zones, even in the absence of competition in a given conflict, is associated with shorter wars more so than in cases when privately held companies offer their services to struggling governments. Our goal is to establish conditions under which PMSCs' accountability could be maximized in ways that could pave the road for quicker termination of wars.

The concluding chapter revisits the main argument and findings about the significance of market pressure on PMSCs' effectiveness in civil wars and thus on their contribution to terminating the violence. We focus on policymaking implications for those interested in ways to improve the monitoring of private military and security industry. Our ideas about the role of market pressure in increasing accountability build on existing efforts by governments and NGOs to develop stronger mechanisms of transparency and adherence with humanitarian laws among private security providers. While many initiatives have been undertaken in this direction, third-party monitoring, an integral aspect of these undertakings, remains challenging to implement on the field. Therefore, innovative and effective approaches toward increasing accountability of the industry continue to matter. The states' ability to recognize and take advantage of the growing and changing competitive environment in the industry may increase the odds that private contractors will align their interests with those of the state, and increase the power of the state vis-à-vis the rebels. We thus present an alternative, market-based, regulatory framework that takes existing approaches in a new direction.

Lastly, we offer ideas for expanding existing research on PMSCs and their ever-evolving impact on security around the world, and discuss practical suggestions for policymakers interested in maximizing the use of PMSCs. Our general recommendation is for scholars to consider the different ways in which private military and security providers adapt to the changing environments in which they operate. Insights from our findings also suggest that specific conflict complexity may constrain contractors' effectiveness even if accountability is high. As such, governments should also develop realistic expectations of what contractors can accomplish in light of the changing conflict characteristics. Furthermore, policymakers play a significant role in improving PMSCs' effectiveness because market dynamics can only work if the client chooses to punish companies that are not militarily effective and reward those that are. When this process is broken, competitive forces become less effective in improving PMSCs' accountability and effectiveness.

# 2

# Private Military and Security Industry

## Opportunities and Challenges in Shaping Conflict Dynamics

The end of the Cold War has been ubiquitously associated with changes in the security landscape. As the Soviet-US rivalry dissipated, traditional alliances between the two superpowers and their pawns in Africa, Asia, and the Middle East have loosened. Unable to count on the rival superpowers' support to strengthen their weak armies, governments around the world have created a demand for professional services that would help bolster their flagging armies in the midst of domestic instability and rebel incursions. At the same time, the end of the Cold War heralded a new development: the growing presence of out-of-work military men faced with realities of cuts in US defense budgets. Skilled, well-connected, and sensing the market's demand for professional military services, many of such men either joined or established PMSCs. While mercenaries have been around since ancient civilizations, PMSCs have emerged as a new breed of actors. Unlike mercenaries lacking a permanent basis from which to operate and whose activities are outlawed under the 2001 UN International Convention against the Recruitment, Use, Financing, and Training of Mercenaries,[1] PMSCs are legally recognized corporations with permanent structures. Individuals working for a PMSC are hired by a private firm, which then receives contracts from governments (mostly) or rebel organizations (rarely). Thus employees working for a PMSC are subject to the firm's regulations in ways that impose greater scrutiny of their performance. Working for a PMSC can also be a source of high and stable income, whereas soldiers choosing to bypass international law and working on their own must constantly negotiate and forge new relationships with potential clients and operate underground.

The market demand for skilled military and security services thus is part of the story of PMSCs' rise; the other is inexorably linked to the availability of qualified and interested men joining the ranks of the PMSC.[2] As the demand for PMSC services has increased, the market has become saturated with

*Private Militaries and the Security Industry in Civil Wars.* Seden Akcinaroglu and Elizabeth Radziszewski, Oxford University Press (2020). © Oxford University Press. DOI: 10.1093/oso/9780197520802.001.0001.

corporate military players. The growing presence of these non-state actors in conflicts around the world has led to a number of questions. Can PMSCs meet the expectations of their clients in ways that could help reduce the duration of conflicts in failed states? Can a corporation interested in profit refrain from selling short-term solutions to clients that, while successful in winning a battle, could jeopardize the government's ability to win the war? Finally, what kind of control mechanisms could be implemented to improve PMSCs' effectiveness in war zones and raise the bar of accountability? Put simply, how can we better understand conditions under which PMSCs could become more effective in ending violence in civil wars?

## The Costs and Benefits of Relying on PMSCs in Conflicts

Existing approaches that attempt to tackle questions related to PMSCs' effectiveness focus more extensively on the negative consequences of such actors' presence in civil wars and ultimately suggest a need for aggressive mechanisms of accountability. Unlike public forces, PMSCs deliver services for profit and a majority of their employees cite financial gain, as opposed to ideology or national interest, as the motivating factor for joining a company.[3] Thus, even if they are "tamer" than freelancing mercenaries, their existence as private firms could still create problems with accountability in conflicts.[4] When financial gains top the agenda, problems with overcharging the government begin to surface. In 2011, for example, a congressional study estimated that the amount wasted in payments to private contractors in Iraq and Afghanistan amounted to anywhere between $31–60 billion.[5] Without proper monitoring mechanisms in place, some PMSCs may purposely operate at lower levels of effectiveness to keep the conflict ongoing and demand additional resources.[6] Considering that PMSCs are among the leading experts in military and security affairs, it becomes difficult for the governments, especially weak ones, to determine whether a need for additional resources is necessary to respond to a complex conflict environment or exaggerated to extend the contract. The net result is that companies can underperform to extract additional payments.

The PMSCs' corporate identity is also linked to other types of behaviors that could be detrimental to the termination of civil wars. Concerns for avoiding risks are greater among PMSC employees than among public forces such as the US military.[7] Reluctance to engage in dangerous operations could inspire rebel groups to continue the fight rather than negotiate with the government, thereby prolonging the fighting. Even if private contractors stay, their goals do not always reflect those of the governments that hire them. Some companies may support opposite groups in conflict, thereby exacerbating the tensions. On rare occasions, the profit-driven interest of private contractors could make them potential clients of both legitimate and illegitimate players, including rebel groups and terrorist organizations,[8] though this is less likely the case for international PMSCs. Consequently, rebel groups with access to funds from drug trafficking or the sale of precious gems could secure sophisticated technology and weapons from private contractors and acquire the means to overthrow the government or, at the very least, wage long wars. In extreme cases, quasi-security companies have deliberately undermined security in a state to coerce the government into seeking their services. Such has been the case in Afghanistan, for example, where warlords have formed security companies while concurrently using militias to maintain tensions in the region.[9]

Finally, there is an issue with PMSCs' effectiveness in cases when corporate actors work alongside the military. Relying on data from interviews with US military and PMSC employees, Dunigan (2011)[10] shows that structural and identity problems have been rampant among private security contractors and the US military fighting together in Iraq and Afghanistan. Differences in structural factors including capabilities, hierarchical structure and order, and in issues concerning identity-related factors such as constitutive norms and social purpose have negatively affected coordination between these seemingly different actors. Gaps in coordination, Dunigan (2011)[11] argues, have resulted in the weakening of the military's responsiveness in conflict. It would not be uncommon, for example, for a military unit to come to the rescue of private contractors without prior knowledge of such contractors' activities in the field.

The negative focus on PMSCs' presence in conflicts is not merely limited to their impact on termination of conflicts. Petersohn (2017)[12] shows that when PMSCs provide combat-related services and face incentives to improve military effectiveness, their intervention expands military activities and thus increases war casualties. In the context of Iraq, lack of contractual specifications and intra-service competition has limited PMSCs' accountability and increased the likelihood of violence.[13] Furthermore, PMSCs' presence has also been linked to long-term instability in weak states after the fighting ends.[14] For example, compensating PMSCs with mining contracts instead of cash has curtailed governments' control of major assets that are vital to generating national wealth in post-conflict reconstruction. Yet increasingly, cash-stricken governments sell mining and oil rights in exchange for security while potentially neglecting the adverse consequences of such transactions on economic growth.[15] Short-term security incentives are also behind the decision to hire PMSCs as the guardians of public security. While this may seem beneficial in states where lack of professionalism among the military threatens stability, Leander (2005)[16] argues that excessive reliance on private security firms diverts funding from improving public armies in the long run. This, in turn, pushes soldiers either to join commercial activities or to contest the status quo by siding with the rebels, the most notable case of this situation being Sierra Leone, where soldiers deprived of cash blurred the distinction between public and private security, working for the government by day and for the rebels by night.[17]

Despite the myriad of problems associated with PMSCs' presence in conflict zones, the provision of private security and military services is a booming business. Companies such as G4S and DynCorp are corporate giants[18] that have delivered unprecedented benefits to governments that rely on their services to bolster their army, provide security, protect strategic resources from rebel attacks, and provide cutting-edge technologies. The use of PMSCs in conflict appears as a cost-effective strategy designed to restore order in war zones. The demise of the Soviet threat and the end of the Cold War have greatly reduced great-power interests to intervene in weak states plagued by years of domestic unrest. The dramatic failure of US intervention in Somalia in 1993 has sealed the fate of such missions—no longer would the public tolerate risky foreign adventures in remote parts of the world where national interest appeared limited. Yet leaving such troubled spots unattended could present serious security risks. Weak states such as Yemen and Somalia, for example, are a refuge to Al-Qaeda. Private military firms could thus fill the security vacuum left by great powers and bolster weak governments[19] either by being directly hired by such governments or by being employed to advance US interests without incurring major public costs. As public accountability associated with using PMSCs in conflict zones is lower than the scrutiny that democratic governments face for sending public forces abroad, reliance on private contractors in conflict zones makes them particularly attractive tools of foreign policy. According to Tim Spicer, chief executive of Aegis, "the impact of casualties is much more significant if they're sovereign forces as opposed to contractors."[20]

With sophisticated weapons and flexibility, these companies, some scholars argue, effectively replace ill-equipped and poorly trained government armies and help terminate conflicts.[21] They are also more skilled and easier to mobilize than most UN peacekeepers, which makes them useful for safeguarding peace and complementing international efforts to resolve conflicts.[22] Although PMSCs are not currently used in the frontline of peace operations, if hired as peacekeepers, they could deliver services at a competitive price. Executive Outcomes, the South African PMSC, charged the government of Sierra Leone $35 million for a period of 21 months, whereas the presence of UN peacekeepers for eight months would have cost a staggering $47 million.[23] Not surprisingly, then, Brooks (2000)[24] argues that in the near future the UN will no longer be able to disregard the benefits that such armies offer in peace-stabilizing operations. Considering the growing demand for peacekeepers around the globe and the UN's inability to send personnel to all the places where the need exists, turning to private security forces could be an effective substitute for traditional peacekeeping and provide PMSCs with new responsibilities beyond the current secondary role of guard protection, risk assessment, strategy development, and support for humanitarian activities.[25] The other side of the story surrounding the debate about the impact of PMSCs on conflict dynamics is thus the positive impact of private military industry on terminating violence by strengthening weak governments and working to ensure the survival of peace after the fighting ends.

## Existing Approaches to Improving Accountability

While scholars continue to ponder the extent to which PMSCs improve security in conflicts or breed instability, most recognize the undeniable fact: the industry is here to stay. Consequently, there has been a growing emphasis on contemplating solutions that would aim to increase accountability among security providers and make them more effective in meeting the clients' interests. Existing solutions to increasing PMSCs' accountability usually focus on greater regulatory efforts at the international level, domestic attempts on the part of the state to introduce more control over the industry through licensing and contract clarity, and the reliance on market forces that encourage self-regulation. Continuous emphasis on ideas and efforts to address the industry's accountability in armed conflict demonstrates the seriousness of such concerns among scholars, NGOs, and policymakers.

Significant improvements have been made to regulate the industry, although the regulatory process exhibits some loopholes. Because the process has flaws, innovative approaches designed to improve accountability in the private military and security industry continue to be of interest. This is especially important in light of the continuous presence of such actors in civil wars across the globe.

At the international level, governmental experts led by Switzerland and the International Red Cross developed the 2008 Montreux Document. This initiative brought together 17 governments and relied on insights from industry representatives, academics, and NGOs to clarify international legal obligations of states related to the use of PMSCs in armed conflicts. While not legally binding, the Montreux Document has been signed by almost 50 states. It delineates 70 good practices that states can embrace to meet the standards of international humanitarian law, offering suggestions for the establishment of an authorization system for military/security services abroad, the criteria that should be used when states consider hiring private contractors, and ideas for strengthening states' monitoring capacity.[26] While the document is addressed to states, it highlights expectations for PMSCs as well. For example, in Section E of Part 1, entitled "PMSCs and their Personnel," the document notes that PMSCs are "obligated to comply with international humanitarian law or human rights law imposed upon them by applicable national law" and notes in the latter part of the same section a specific reference to PMSC personnel who are "obliged to comply with applicable international humanitarian law."[27] By setting clear guidelines on respecting international humanitarian law and delineating criteria for the selection of PMSCs by states, the document indirectly communicates to the industry what the clients could be looking for in the future, therefore putting pressure on companies to respond accordingly. The practices are addressed to contracting states or those that seek the services of a PMSC, home states of the nationality of a PMSC, and territorial states or those on whose territory PMSCs operate. Recommendations for strengthening the states' monitoring capacity range from establishing broader mechanisms of criminal accountability such as ensuring criminal jurisdiction in national legislation and jurisdiction for crimes committed by PMSCs abroad to more specific non-criminal accountability mechanisms such as terminating companies' contracts or imposing financial penalties in the form of reparations to victims of PMSCs' misconduct.[28]

The focus on setting specific standards that would improve PMSCs' accountability has continued with the ANSI/ASIS International's 2012 Management System for Quality of Private Security Company Operations, which was updated in 2017, and ISO's 2015 Management System for Private Security Operations 18788. Both initiatives implement the Montreux Document's recommendations and build upon the industry-managed International Code of Conduct for Private Security Service Providers' (ICoC) efforts by developing risk-assessment criteria that companies can meet to show their commitment to improving practices that would reduce the risk of human rights abuses in their day-to-day operations.[29] Companies evaluate their practices by developing assessment criteria, testing, reports, and providing reflections on lessons learned. After going through an internal audit, they seek certification by an independent third party. The US and UK, major consumers of the industry's services, now require either ANSI/ASIS or ISO certifications.[30]

At the domestic level even before the Montreux Document, ANSI/ASIS, and ISO initiatives, governments sought to increase PMSCs' accountability through a licensing system. A strong national regulatory framework was to ensure that PMSCs working in a conflict zone would have the proper experience and resources to conduct the mission successfully. This would be accomplished through an efficient licensing system that would evaluate the companies' capabilities and determine if they could operate abroad.[31] For example, the Arms Export Control Act in the US requires a license for a PMSC that intends to provide defense services to foreign governments.[32] The licensing system in general, however, has a major limitation. Gaining a license determines whether a given company has the capacity to fulfill the contract but is not automatically designed to monitor the companies' behavior in a conflict zone. The latter task ultimately falls in the hands of the governments and specific agencies that award the contract.

In the context of the US, there have been differences in how agencies have addressed licensing and monitoring issues in Iraq. Initial monitoring efforts have been somewhat more robust for contracts with the Department of Defense (DoD) than the Department of State, although both have struggled with shortages of personnel to strengthen oversight. The Department of State has exempted some of its clients from US military regulations regarding licensing, as was the case with Blackwater in 2007. Although the DoD made it mandatory for a security company to obtain a license from the Iraqi Interior Ministry to operate legally in Iraq, Blackwater, which received contracts worth $678 million, was free to roam the country without one.[33] Companies that have received their contracts through the DoD abide by US military regulations that place limits on the use of offensive weapons—and thus could limit human rights abuses—and are subject to a tracking system that monitors private security providers' movement on the battlefield.[34] Additionally, while the DoD participated in processing complaints regarding coalition PMSCs through the Reconstruction Operations Center, which became operational in Iraq in 2004 to facilitate military and contractors interactions, the Department of State did not.[35] Still, both agencies faced challenges with monitoring contractors' performance in the field. In the past, US embassy officials have been tasked with overseeing fulfillment of services,[36] while at other times, the Department of State designated contracting officers to meet with private security contractors in a conflict zone and Washington, DC, and coordinate with other federal agencies to ensure compliance with regulations.[37] Contracting officers have also been tasked with monitoring contractors' activities for DoD contracts.[38] Ensuring compliance with regulations was difficult when the contract was awarded by the Department of State, as contracting officers or embassy staff have not always accompanied PMSC employees on the missions.[39] In its 2009 report, the Commission on Wartime Contracting in Iraq and Afghanistan found that all federal agencies responsible for contracting experienced shortages of qualified personnel to check contractors' performance as the presence of contractors increased.[40] Overall, different, at times incomplete, approaches toward licensing and monitoring of the companies' activities on the field highlight gaps in domestic efforts to increase the industry's accountability.

Problems with licensing and monitoring have existed in other contexts as well. South Africa has attempted to regulate the industry through the Foreign Military Assistance Act of 1998 and through the 2006 Prohibition of Mercenaries Activities and Regulation of Certain Activities in Country of Armed Conflict Act.[41] The Foreign Military Assistance Act required authorization for PMSCs to start negotiations with a potential client and governmental approval of the contract. Despite a seemingly comprehensive attempt to increase accountability, this measure has been mostly futile in preventing companies from seeking clients without authorization. Lack of enforcement and small penalties faced by companies choosing not to seek contract authorization have created limited incentives for PMSCs to abide by the rules.[42]

Improvements in specific contractual structures are a promising tool for increasing PMSCs' accountability but these efforts have been problematic, as is evident in the context of Iraq. Dickinson (2007)[43] points to the unprecedented vagueness in Iraqi military and reconstruction contracts that the US entered into when compared to contracts that local and state governments issue, for example, to manage prisons. "Foreign contracts possess so few guidelines, requirements, or benchmarks that they effectively contain no meaningful evaluation criteria whatsoever."[44] When Pacific Architects and Engineers (PAE) was hired by the US and Afghan governments to support operations of the Afghan counternarcotics unit from 2007 to 2010, the company performed services even when its proposed costs for task completion were awaiting approval. This and a lack of set standards against which PAE's performance would be measured meant that the clients could not be sure that they received high-quality services at a fair cost.[45] Difficulty in gathering accurate information about the cost of services in conflicts also means that clients suffer from an informational disadvantage when it comes to pricing estimates, enabling cost inflation by PMSCs.[46]

In addition to displaying vagueness, contracts during the early stages of the war have been notorious for missing requirements to comply with international human rights laws by service providers. Dickinson (2007),[47] who analyzed all publicly available contracts for military and reconstruction operations entered into by the US government prior to 2007, noted that none required PMSCs to demonstrate respect for human rights, transparency, and anticorruption norms. She argued that accountability could be improved by developing contracts that require contractor training in the area of international human rights law, creating specific performance benchmarks, and demanding accreditation of contractors by independent organizations. Inclusion of legal social responsibility considerations in contracts awarded in Iraq, Afghanistan, and beyond has improved since, especially in the area of accreditation with ANSI/ASIS or ISO standards.

While these improvements offer benefits in increasing contractors' accountability, two main problems persist. First, even though more specific performance benchmarks have been included in contracts that could limit criminal behavior, the practice is not uniform[48] and is likely to be less institutionalized among clients that fall into the category of weak states. Furthermore, specific benchmarks might not always be optimal. Some degree of flexibility might be beneficial in cases when PMSCs are in a better position than their clients to assess the needs while in the field. Yet this again presents an opportunity for some PMSCs to subvert the client's interests by redefining contract terms, convincing states of the need for additional services or justifying a delay in effectiveness by citing evolving complexity in a conflict environment. Second, despite greater emphasis on compliance with human rights laws in contracts, there is no guarantee that companies will abide by these standards once they receive the contract and are operating in the field. Just like with licensing to operate abroad and ANSI/ASIS and ISO certifications, effective monitoring becomes difficult to accomplish when contractors are out on a mission in a war zone and operate alone without a public army by their side[49] or with limited presence of contracting officers. Overall, what this suggests is that improvements in contract quality do not automatically ensure that military and security providers will be accountable to the client.

Increasingly, scholars have focused on the role of market forces, especially the growing competition among PMSCs, in making private military and security providers more accountable. The PMSC market is a crowded one; while in 1989 there were 15 providers, 300 companies delivered a wide range of services in 2006. Clients have options to select companies based not only on the cost and type of services they provide but also on their reputation for effectiveness and accountability. As PMSCs are profit-oriented actors, they are thus most likely to respond to market pressure; their need to secure clients should push them to self-monitor their behavior in conflict zones in ways that would maximize their effectiveness and communicate to the client their interest in accountability. Even prior to ANSI/ASIS and ISO certification and in addition to it, many PMSCs have joined trade associations upholding ethical standards to demonstrate commitment to greater professionalism.[50] Some have adopted stricter rules against behaviors that might damage the companies' reputation for accountability. According to one Blackwater employee, for example, the company implemented a zero-tolerance policy for its employees after the events in Iraq and Afghanistan tainted the company's image. Any suspicion of behavior that went against international humanitarian standards or involved fraud and corruption, for example, would have resulted in immediate dismissal of an employee, with limited or no investigation of accusations.[51] Stricter penalties may then create an incentive among employees in war zones to monitor their own actions or risk dismissal. The effectiveness of such a monitoring mechanism, however, is questionable. According to a former Blackwater employee, it would not be uncommon for cases of poor behavior to go unreported to the corporate office. At times, the leader of the group that was sent on a mission would issue a warning but immediate penalties would be rarely implemented.[52]

The most notable example of PMSCs' efforts to establish good reputation in a crowded, competitive marketplace is evident in the large number of companies that have signed the International Code of Conduct for Private Security Service Providers (ICoC), which was finalized in 2010. Initiated by Switzerland with cooperation from civil society organizations, PMSCs, academics, and other states, the Code is an initiative designed to set out and clarify rules for responsible delivery of services by private security companies.[53] The rules, adopted by over 700 companies, touch upon such areas as the vetting of personnel, the use of force, and prohibitions against human trafficking, torture, and human rights abuses. The signatories also agree to the establishment of a Steering Committee whose task is to develop initiatives for independent monitoring of the industry. Once such mechanisms are in place and companies submit to an independent auditing and monitoring system that verifies adherence to the code, the companies are granted the Certification of Signatory Companies' compliance with the code. As part of the certification process, companies conduct internal reporting on their performance and note the changes that have been made to meet the Code's principles. In addition, companies must submit to auditing in the field by a third party.[54]

The oversight mechanism of ICoC has the potential to increase accountability among private security providers in ways that could make them more effective actors in terminating conflicts. The on-the-field monitoring component addresses some of the earlier problems associated with issuing licenses, specifically the problem of not being able to verify what happens in the conflict zone once the companies obtain the license and become participants in the conflict. Overall, the oversight mechanism embedded in the ICoC represents an ambitious and comprehensive attempt to monitor the industry. The certification process determines whether companies have internal mechanisms and structures in place to meet the Code's principles and then verifies that good practices continue in the field. It is the latter aspect of PMSCs' accountability that has been notoriously difficult to address. At the same time, it is difficult to assess how the ICoC's oversight mechanism works in practice. For example, determining accountability in the field requires that independent observers are present in some of the most dangerous conflict areas over extended periods of time. It is not clear whether independent third parties will have the willingness and resources to engage in more in-depth accountability assessment. The ICoC's monitoring mechanism is highly dependent on reports from civic organizations, whistleblowers, and industry members to reveal Code violations, thereby highlighting the importance of such "informal" sources in boosting accountability. Yet the commitment to follow up on informal tips regarding possible Code violations is not guaranteed, as the Association that issues the code relies heavily on contractors' membership fees to function.

There is also a number of specific trade organizations that PMSCs have joined to communicate their professionalism and support for the codes of conduct these organizations have developed. As members of the International Stability Operations Association (ISOA), the British Association of Private Security Companies (BAPSC) until 2011, or the Private Security Company Association of Iraq (PSCAI) until 2011, companies received information about ways they could improve accountability and were asked to honor principles related to, for example, human rights protection, accountability, and ethics.[55] In Iraq, the PSCAI worked with coalition governments and the Iraqi government to develop guidelines for practices that would align private contractors' work with Iraqi laws and respect for Iraqi communities until the organization was disbanded in 2011.[56] These attempts to self-regulate, however, have had notable limitations in credibly communicating commitment to accountability. The ISOA's enforcement mechanism, for example, allows anyone to file a complaint against the company, which is then evaluated by the association's Standards Committee. Yet the consequences are not severe enough to generate reputational costs that might merit future loss of contracts. The Standards Committee can recommend policy changes upon hearing a complaint, place a company on probation, or expel it from the association. As complaints against a company are not made public,[57] unless the Committee expels a PMSC from the association—a rare possibility—there is a notable absence of negative publicity for the company that could potentially result in a cascading effect of client loss over time. Put simply, the reputational costs from the Committee's negative ruling are unlikely to change the companies' behavioral incentives and increase self-regulation. As such, the need to develop additional accountability mechanisms continues to be an issue of interest to scholars and policymakers.

## Market Dynamics: Competition, Accountability, and Military Effectiveness of PMSCs

We build on the literature that highlights previous and current initiatives designed to improve PMSCs' accountability by unpacking specific market mechanisms that could increase military effectiveness of PMSCs and make the companies more accountable to the clients' needs in civil wars. While an increase in competition among security providers on a global scale has put greater pressure on PMSCs to increase their accountability to the client, private companies also face pressure from what we refer to as *local or conflict-level competition*: the presence of multiple security providers in a specific conflict. While most companies may be exposed to similar levels of competition stemming from the industry's expansion, there will be considerable variation in local competition: in some conflicts, a government may work with only a single provider, but in others, contracts are awarded to multiple players. Companies facing global and also local competition experience more pressure to be effective. They must bid successfully to offset global competition for a contract, and then demonstrate military effectiveness in the field to highlight their success in future bids. And so the competition does not immediately end once the contract is awarded but continues in the field, especially in those contexts where multiple companies operate in a given conflict zone. In his work on the regulation of private security forces, Jeffrey Herbst (1999)[58] highlights the impact of competition on PMSCs' behavior, noting that "in industries where the barriers to entry are low and where, as a result, companies probably cannot compete on price alone, firms will necessarily attempt to differentiate themselves in other ways."

The presence of local competition creates a mechanism for improving military effectiveness. Numerous definitions of military effectiveness exist, from reference to a mission fulfilled in a combat zone, to the readiness of forces to be used in wars.[59] We rely on Newell and Sheehy's (2006)[60] definition of PMSCs' *military effectiveness* as improving the client's ability to accomplish its military mission. We refer to *accountability* as the extent to which PMSCs are responsive to the client.[61] PMSCs that exhibit a high level of military effectiveness are likely to be highly accountable to the client. High level of accountability, in turn, should be associated with greater *shifts in the balance of power* in favor of the government and higher odds of civil war termination. The realist school of thought argues that shifts in the balance of power occur when one side gains advantage in military capabilities over another that results in an outright victory or ability to compel the other side to negotiate by increasing the costs of fighting.[62]

Military capabilities include not only weapons and personnel but also training, logistics, intelligence, and communication. With their wide-ranging expertise in these areas, PMSCs can increase the government's military capabilities to help the latter track and kill insurgents. For example, the use of more sophisticated equipment such as night-vision goggles and infrared lasers by Afghan commanders has been useful in fighting insurgents at night.[63] PMSCs can also help train the army in military tactics designed to formulate a long-term strategy of defeating the enemy and in asymmetric conflicts such as insurgencies. In asymmetric conflicts, insurgents, whose equipment and manpower are inferior to the government's, will depend heavily on the population's support for intelligence, resources, and recruits. In such contexts, altering the balance of power in favor of the government involves not only enemy killing but also gaining support of the locals to limit their assistance to the insurgents. The idea in such contexts is to change the perceptions of the population to see the government, and not the insurgents, as a reliable and legitimate guardian of their interests. In the most fundamental way these interests concern security. In a war zone the government and the insurgents compete with each other to demonstrate who can better provide this basic resource to the population.[64] Insurgent forces often exploit vulnerability of the locals, in hopes of turning the people against the government to increase their loyalty and perceptions of the insurgents as alternative security providers. By hiding among them and creating conditions that increase the odds of government attacks harming the civilians, the insurgents can exploit the insecurity to drive more individuals to join their cause.[65] PMSCs can help governments in this area by training the forces to conduct counterinsurgency operations that reflect sensitivity toward civilians in coordinated or individual operations and encourage restraint in the types of weapons they use in the field to minimize indiscriminate killing.[66] Yet to improve the effectiveness of such efforts, PMSCs must be willing not only to promote such strategies as part of the training portfolio when given a chance but also to adhere to the very principles of humanitarian treatment of civilians themselves when completing the tasks they have been assigned.[67]

In the subsequent chapter we investigate how local competition improves monitoring of PMSCs' behavior in conflicts and thus creates willingness for such players to become more militarily effective and, in turn, more accountable to the client. Furthermore, we argue that in response to *global competition* or the growing competition at the industry level, some companies have adopted a unique, more transparent corporate structure that has resulted in more aggressive self-monitoring with the potential to improve the companies' effectiveness. While gaining capital is the primary goal behind a decision to go public, this decision has a direct impact on improving transparency and accountability due to more stringent legal requirements regarding information sharing about publicly traded companies' activities with the public. Becoming a publicly traded company signals a degree of uniqueness among the ever-growing pool of private contractors. As clients are not always able to clearly distinguish among different security providers' quality of services,[68] a transparent *corporate structure*, or the way in which the business practice is organized, can help the client gauge the level of military effectiveness they can expect from the company. With the public and the media serving as the watchdog, publicly traded PMSCs or companies with transparent corporate structure face a much greater level of scrutiny than do companies that are privately held. The net effect of this dynamic is a greater likelihood of effectiveness in conflict zones and greater accountability to the client. Once the companies are highly accountable, the balance of power is more likely to shift in favor of the government and contribute to shorter conflicts. Figure 2.1 captures our argument about the relationship between market pressure, military effectiveness, PMSCs' accountability, and civil war termination.

In Chapter 3, we rely on principal-agent logic to revisit the problem of military effectiveness for PMSCs in civil wars. Yet before we do that we highlight specific components of military effectiveness where PMSCs can make a difference. While scholars have turned to several dimensions of military

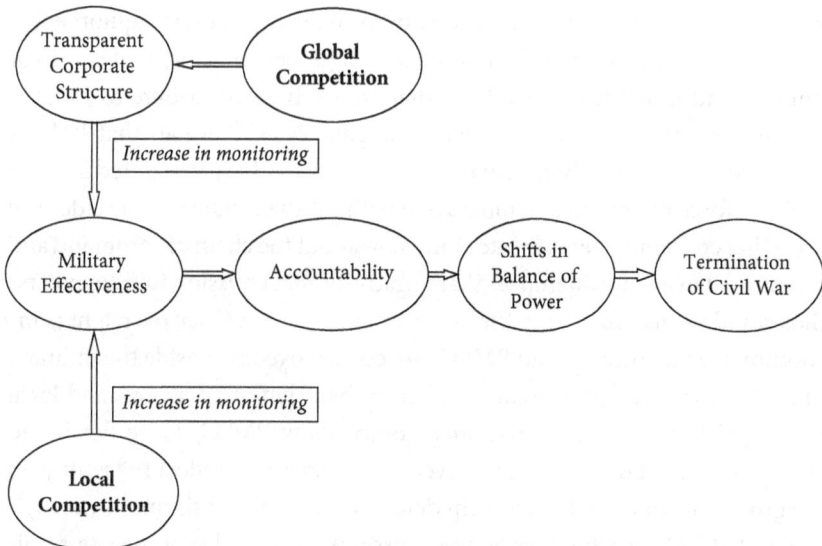

Figure 2.1  Market dynamics, PMSCs' accountability, and civil war termination

effectiveness, our focus is on two elements that are critical to PMSCs' fulfillment of the mission in ways that are consistent with client expectations of
meeting their interests in a conflict. These elements include more-traditional
elements emphasized by neorealist scholars as well as those developed by
constructivists that emphasize norms and cultural practices. Ultimately,
our emphasis reflects the idea put forward by Millet, Murray, and Watman
(1986)[69] that measurement of military effectiveness depends on specific
conditions faced in a given context. For example, in the context of many civil
wars, which are the subject of our investigation, adherence to human rights
laws may be more important in achieving military effectiveness than in international conflict, as we will subsequently argue.

First, we focus broadly on material and individual capabilities that could
increase PMSCs' military effectiveness. Key indicators of such capabilities are
noted by Brooks (2007)[70] in the context of the military; Dunigan (2011),[71]
who looks at PMSCs' effectiveness in Iraq; and Fitzsimmons (2013),[72]
who examines mercenaries' and PMSCs' effectiveness in four asymmetric
conflicts. These include: the integration of military/security activity within
and across different levels,[73] for example, structural integration into the military force alongside which PMSCs are delivering services;[74] responsiveness
and adaptability, which denotes the ability to react to the enemy's capabili-
ties/external environment[75] and flexibility in responding to evolving conflict
dynamics;[76] skill, which is motivation and competencies of personnel[77] as
well as creative thinking and the ability to maintain military equipment;[78]
and quality of weapons and equipment.[79] Together these attributes mirror
military and individual capabilities that are likely to contribute to PMSCs'
military effectiveness. For example, Dunigan (2011)[80] argues that PMSCs
would be more militarily effective if they were either fully integrated into the
military force or sent to a combat zone without the military by its side. She
finds that co-deployments create confusion about the chain of command and
resentment over pay differences that negatively affect mission fulfillment and
thus could be detrimental to the government's victory. Once oversight is institutionalized, however, and PMSCs are co-deployed alongside the military,
in many cases their performance is better than that of US troops and local
military.[81] Responsiveness and adaptability allow PMSCs to be flexible to
changing conflict dynamics and develop strategies that reflect the reality on
the ground in ways that could help defeat the enemy. Fitzsimmons (2013)[82]
shows that PMSCs whose employees possessed knowledge of the latest military technology, displayed the ability to maintain military equipment, and

demonstrated innovative thinking during critical moments of the mission were able to gain substantial advantage against the enemy. Finally, Dunigan (2011)[83] relies on interviews with military personnel in Iraq to argue that PMSCs' ability to provide high-quality weapons and equipment was seen as a contributing factor to the success of US operations.[84]

Second, we consider corporate professionalism to be an important factor of PMSCs' military effectiveness. Corporate professionalism refers to the socially acceptable core values that define the mission of the company and are expected to be applied by its employees.[85] For public forces in Western countries, such as the United States, the idea of professionalism is firmly ingrained and institutionalized. Therefore core values that denote professionalism, such as integrity, subordination of the military to the democratic civilian authority, and allegiance to the state, are clearly defined with rules to penalize behavior that deviates from those values.[86] Such values flourish in organizations that have a clear mission and are accountable. Public forces across the world exhibit different levels of professionalism. In the context of PMSCs, as we will argue in Chapter 3, there is also a tremendous variation in the commitment that an organization places on professionalism and this disparity can, to some extent, be explained by the variation in competitive pressure that companies face. While allegiance to a state falls outside of the PMSCs' domain, the value of integrity in particular is relevant for the industry, and thus we focus on this element of corporate professionalism in greater depth. Integrity denotes behavior that is in accordance with the standards set by the company, institution, or the profession as a whole rather than guided by personal interest.[87] In the case of PMSCs these standards have been laid out by the profession in conjunction with scholars, NGOs, and policymakers through initiatives such as the Montreux Document and ICoC. Individual clients also set expectations for security providers in contracts, for example, by expecting accreditation with ANSI/ASIS or ISO Standards. When we refer to integrity we thus consider the organization's commitment to respecting international humanitarian law and honest business practices that focus on meeting the interests of the clients and the key values emphasized by the industry, governments, and non-governmental organizations.

## Reducing Fraud and Military Effectiveness

How is corporate professionalism relevant for military effectiveness? Consider the first dimension of corporate professionalism that we focus on, honest business practices that reduce the occurrence of fraud. Fraudulent practice in the case of PMSCs occurs when there is a deliberate intent to abuse the clients' interests for personal gain. If clients are deliberately overbilled for ghost services that never exist, companies use a different type of equipment in conflict than promised, or officials are "sold" on additional yet unnecessary services, then resources are extracted from a government that could be spent to hire another PMSC to add value to the mission. While wasteful spending is unavoidable and to some extent expected, it is bound to have a negative impact on all states, with bigger consequences for weak states that already possess limited capacity to tax citizens and generate extra revenue. Cash and resources are not readily available to all governments, and overbilling a client for services could mean that the government has fewer resources to accomplish a mission that might play an important role in conflict termination.

Even in stronger states, there is a high likelihood that saved resources would be used to strengthen existing programs rather than transferred to completely different policy areas. In the US, reprogramming or the shift of funds within an appropriations account is permitted based on an informal understanding and often a necessary response to changing needs.[88] Thus, money saved from overbilling by one PMSC in the area of base operations support could be used to fund additional PMSC missions in the same category to enhance the service's quality. Transfers or shifts of budgetary resources from one appropriations fund to another require congressional approval. While the DoD does not view congressional approval as legally binding, such approvals are seen as "practically" binding given that violation of trust could affect how Congress approaches the next appropriations process.[89] While it is not impossible to make the transfer, the process is more complex if it involves moving funds to a completely different program, as was evident by the battle between President Trump and Congress regarding the transfer of funds from a military base project to the construction of the US-Mexico border wall. The process ultimately ended with the Supreme Court affirming the power of the executive to reassign funds under a declared state of emergency. Overall, it less likely that saved resources from PMSCs' overbilling in one appropriations account would be easily diverted to vastly different and unrelated policy areas.[90] Restrictions on moving funds are also in place outside of the US context, and are more stringent when the move involves different policy areas or different ministries, for example, a transfer from a country's ministry of defense to its ministry of education. Movement of resources in such cases is either prohibited, severely restricted, or requires approval of the legislature, while movement of resources within the same ministry or department faces some restrictions, albeit smaller, when it involves different categories.[91] What this shows is that saved resources for a specific mission would have most likely ended up being used to enhance that mission, for example, by funding extra personnel, awarding contracts to more expensive but more transparent service providers, or paying for additional contracting officers to improve contractors' monitoring. They could have been also used to fund other, related missions.

Consequently, the effectiveness of PMSCs in Iraq might have been greater had the contractors delivered what they promised and refrained from overcharging the government. In fact, the connection between fraud and military effectiveness in Iraq and Afghanistan is clearly articulated by Schwartz and Church (2013)[92] in their report to Congress on the Department of Defenses' Use of Contractors to Support Military Operations. The authors argue that, "while the total cost of contract fraud . . . may never be known, there is general agreement that billions of dollars squandered . . . could have been used to achieve other operational priorities." Singer (2007)[93] is in agreement: "Such corruption doesn't just represent lost funds; it represents lost opportunities for what those funds could have been used on to actually support the mission: everything from jobs programs to get would-be insurgents off the streets to flak vests and up-armored vehicles for our troops." Similarly, a practice of charging the US government for a specific number of PMSC employees to train Iraqi Emergency Response Unit forces while undersupplying in the field has been associated with undermining US efforts in Iraq.[94] In Afghanistan, DynCorp's failure to provide the agreed number of trainers negatively affected the Afghans' military performance against the Taliban forces.[95] When fraudulent practices have an adverse impact on US operations, there is greater likelihood that it may take longer to defeat the enemy and thus the conflict can be prolonged.

A culture of low integrity is not only problematic when it involves decisions made by the companies' leadership regarding cost and supply of services. Limited focus on honest business practices can trickle down to the employee level and contribute to corrupt practices that may include, for example, accepting bribes from the group that PMSC employees are supposed to defend the government against. PMSCs that do not develop a strong sense of corporate integrity, which institutionalizes ethics and minimizes the use of fraudulent practices, are less likely to develop strict rules against such behavior and credibly communicate a zero-tolerance policy. Even if such rules exist in theory, contractual manipulation at the top level could undermine the seriousness with which lower-level employees approach the issue of integrity in their service delivery. It would most likely reduce the need for employees to report instances when they are put in a position that may push them to give bribes and engage in other corrupt practices. More important, it would increase the frequency of corrupt behavior that could jeopardize the missions in the field.

We recognize that in some instances what may constitute corrupt behavior is not easily visible to a contractor and may not always undermine the war effort. For example, contractors who bribe villagers in exchange for credible information on the location of insurgents might be helping with operations designed to capture the enemy. Corporate professionalism does not imply inflexibility in developing ethical standards, especially in the gray areas; on the contrary, companies with a strong culture of integrity would encourage employees to report ethical dilemmas to the higher-ups rather than foster a culture of opaqueness, indifference, or complicity. It is in the absence of strong professional culture that discussions about what constitutes corruption and when bribe-taking could adversely impact military effectiveness are rarely prioritized. In his analysis of mercenaries' effectiveness in asymmetric conflicts, Fitzsimmons (2013)[96] notes the importance of transmitting accurate military information among different groups. Failure to communicate information and misreporting could jeopardize military missions that might be critical, for example, in helping the government achieve military victory and end the conflict. Professional corporate culture that develops core values of integrity would encourage honest dissemination of information not only among different groups of PMSC employees but also between such groups and the government forces that they might be supporting in the field. When integrity as a business practice is an issue for the company, such honest communication is also likely to suffer.

## Adherence to International Humanitarian Law and Military Effectiveness

The second dimension of corporate professionalism that we argue affects military effectiveness of PMSCs is their adherence to humanitarian law in the context of civilian protection during the conduct of military and security operations. While a considerable variation exists in what constitutes human rights abuses, we focus on the violations against the civilian population and property in the territory where PMSCs operate as outlined in the Fourth Geneva Convention adopted in 1949. These include arbitrary detentions and killings, torture, sexual crimes, and other instances of disproportionate use of force against civilians and property.[97] Respect for humanitarian law is often considered a factor that contributes to military effectiveness in specific types of civil wars, insurgencies, which are defined as protracted violent struggles by non-state actors against existing political authority with the use of guerrilla warfare strategy. The strategies include the reliance on small and mobile groups to carry out hit-and-run strikes and winning the support of some segment of the noncombatant population. Civil wars broadly defined may include direct battles between enemies as well as guerrilla tactics.[98] With the possible exception of the 1992 conflict in Georgia, all of the civil wars in our data set for major wars (1,000 battle deaths) since 1990 are insurgencies. In the minor conflict data set (25 battle deaths), all but five wars fall into the category of insurgencies.

A debate exists in the literature about employing the best counterinsurgency (COIN) tactics to ensure greatest military effectiveness. Some, for example, advocate the use of enemy-centric approaches with the focus on defeating the enemy through conventional means, while others emphasize a population-centric, the so-called hearts and minds approach that involves establishing control over the population and winning their support through state legitimacy.[99] Human rights protection is often the casualty in an enemy-centric approach as the government employs the necessary and aggressive means to attack and decimate the insurgents' strongholds. Occasionally, an excessive type of COIN involving mass killings of civilians has been used by governments as a strategy in guerrilla wars when conventional tactics fail. By killing civilians, the government can deny the insurgents one of the key sources of their strength: the support from the local population. In its extreme form, this "draining the sea" approach can lead to the extermination of entire populations that support the insurgents[100] and potentially shorten the duration of wars. While most military strategists would refrain from advocating mass killings, there are critics of population-centric COIN who point to the approach's detrimental impact on flexibility and quick action that is sometimes necessary to capture the enemy, not to mention that building state legitimacy is a daunting task.[101] Kilcullen (2007)[102] argues that one approach is not necessarily superior to the other and that the evolving and dynamic nature of insurgencies often requires the use of both as part of a more balanced strategy. Springer (2011)[103] supports this idea by showing how an off-balance strategy in favor of either enemy or population-centric COIN has hurt US efforts in Afghanistan. While we do not argue that adherence to humanitarian law, which is a significant aspect of population-centric COIN, alone is sufficient to achieve military effectiveness in insurgencies, the existence of numerous studies that emphasize the utility of winning the support of the locals through security guarantees suggests that limiting severe human rights abuses is likely to contribute to PMSCs' success in fulfilling their mission on behalf of the government, with potentially positive impact on conflict termination.

Consider the following insights that illustrate the benefits of human rights adherence on military effectiveness and success in terminating the war, as well as findings that demonstrate the disastrous effects on war termination when humanitarian concerns are set aside or deliberately violated through mass killings. Empirical and case study research shows that indiscriminate use of force is a strong predictor of local population joining the insurgency to avenge abuses and killings from government forces.[104] In the midst of grievances against violent tactics, the rebel leaders are in a better position to recruit the locals by exploiting their desire for vengeance.[105] Furthermore, indiscriminate violence helps resolve collective action problem by making non-participation in an insurgency costly. When the state punishes or abuses individuals without distinguishing between the guilty and the innocent, the incentive to join the insurgent organization rather than become a victim of state abuse increases.[106] An indiscriminate violence by the state in such cases generates incentives for rebels to offer protection as a selective good for joining their organization, all of which increases rebel recruitment. Lastly, indiscriminate violence also allows insurgent and opposition leaders to use state brutality as a means of spreading their wider propaganda, which diminishes state support. For example, when Iraq's radical Shia leader Muqtada al-Sadr planned to withdraw from the coalition and plunge the country into civil war, he utilized Blackwater's Nisour Square civilian shooting as a propaganda and recruitment tool. Rather than justify his actions, he focused on the incident, describing it as a "cowardly attack committed by the so-called security company against our people without any justification."[107]

While indiscriminate violence by the government can increase the recruitment of supporters for the insurgents, how does the growing number

of supporters for the rebels shape the duration of war? As rebel recruitment increases and the balance of power shifts away from the government,
the termination of conflict becomes more difficult. A growing number of
supporters and the vast social networks they form may provide the rebels
with shelter, food, information, and most important, willingness to fight
as group identity solidifies.[108] When the balance of power shifts away from
the government, rebel leaders are less likely to negotiate as they may expect an outright victory and thereby continue to fight. Even if the rebels are
open to negotiations, their growing support base may push them to make
greater demands from the government: concessions that governments are
usually unwilling to make unless they see no possibility of victory in the
future.[109] In such cases the rebels will continue to wage their military campaign. Put simply, while gaining the support of the locals could help the
governments secure victory in conflicts and terminate the war quickly,
gaining the support of the population by the insurgents will have an opposite effect on war termination because in the majority of conflicts it is the
government that initially has more power than the rebels. Therefore any
shifts in sources of power in favor of the rebels will lead to greater parity
in power with the government yet not be overwhelming enough to secure
a quick victory.

In a different causal story, Kirschner (2009)[110] attributes atrocities committed by the government against local populations to prolonged war, using
the concept of a commitment problem. The idea of commitment problem in
civil wars links the failure of many peace agreements to the combatants' lack
of credible assurances of each other's ability to uphold the terms of the negotiations.[111] Kirschner (2009)[112] shows that commitment problems increase
when the government commits atrocities against the locals and the group
they support because the rebels are less likely to trust the government to uphold its promises given the low value it attaches to civilian life. As such, the
rebels are less likely to engage in negotiations and more likely to continue the
fight that prolongs the conflict.

The case of civil war in El Salvador (1979–1992) illustrates how indiscriminate violence against civilians can offset the government's gains
against the insurgents in ways that inhibit war termination. Despite considerable risks, the rural population actively supported the Frente Farabundo
Martí para la Liberación Nacional (FMLN). FMLN was able to recruit
supporters from the local population after the government sanctioned the
use of death squads by paramilitaries, irregular units, and Mercenary to

instill fear among the population and to help defeat the insurgents.[113] The scale of state abuses motivated aggrieved peasants to act on moral principles and join FMLN to seek justice.[114] As state-sanctioned violence toward civilians became more pervasive and arbitrary, FMLN's support base increased to between 6,000 and 8,000 active guerrillas, 100,000 part-time militia, and one million sympathizers who provided food, intelligence, and safe haven to the rebels.[115] It was during this time that the FMLN engaged in heavy offenses against the state and consolidated zones of control in the first half of the 1980s. At the same time FMLN was not strong enough to defeat the government, and so the war continued. After the failure of repressive tactics, the government finally began to shift its strategy toward accommodation by embracing a population-centric COIN. It engaged in land reforms and promised institutional changes in the military and governance. Despite notable problems with the reforms, it was the state's decrease in the use of death squads and military massacres that turned out to play a role in curtailing covert support for FMLN in the late 1980s.[116] This shift from coercion to accommodation eventually paved the way for war termination in the early 1990s because it strengthened the government's negotiating position against FMLN.

More comprehensive study by Paul, Clarke, and Grill (2010)[117] of COIN strategies and their impact on shaping governments' victory in 30 insurgencies finds support for the population-centric approach. Looking at tactics such as gaining support of the local population, disrupting insurgents' supplies and financing, and the use of overwhelming force, the authors note that population-centric COIN provides the biggest benefits when the gravity of support for the insurgency centers around the population, and not when the rebels rely solely on external sources for support. The argument in favor of securing the locals' trust is also reflected in a large empirical study of 286 insurgencies from 1800 to 2005 (Lyall and Wilson III 2009).[118] Pondering a surprising pattern in the data, which revealed that governments have won most insurgencies before 1917 yet have struggled to defeat the enemy since, Lyall and Wilson III (2009) contend that in the nineteenth century, armies developed strong connections with the locals, a capital that yielded trust and access to information about the population. With better information, the armies could distinguish between insurgents and combatants to avoid indiscriminate targeting. By contrast, the shift toward mechanization of the armies has diminished connections with the locals and increased the prevalence of collective punishment. Lack of information about the population has compounded the inability to distinguish between combatants and civilians. This dynamic has benefited the insurgents by increasing their ability to recruit among the locals and thus hurt the government's effectiveness in winning the war.[119] Although Lyall and Wilson III (2009)'s study does not specifically examine the duration of war, the likelihood of conflict termination declines when the insurgents are able to recruit followers and the balance of power begins to shift away from the government.

What about the evidence in support of the most extreme COIN tactics on military effectiveness? In the case of repressive and collective punishment or crash-them tactics, success is achieved only in intermediate phases, but such tactics mostly fail in delivering a decisive victory to the government, note Paul, Clark, and Grill (2010)[120] in their study of insurgencies. Since a civil war in our data set is considered terminated when peace is sustained for two years, intermediate victories fail to qualify as termination of war. Overall, as Paul, Clarke, and Grill note, "repression does not guarantee defeat, but it is unambiguously poor approach," as they find only two cases where the governments achieved military victory by utilizing the approach (Turkey and Croatia).[121] Similarly, Valentino, Huth, and Balch-Lindsay (2004) note that although governments do resort to mass killings of civilians, these strategies often backfire. "Mass killing can keep guerrilla forces at bay, but even the most extreme levels of violence are often insufficient to decisively defeat mass-based insurgencies."[122] Thus, repressive tactics are less likely to be considered a factor that increases military effectiveness in terminating the war.

Beyond scholarly findings, governments and leaders of insurgencies, on average, recognize that winning over the population is the path to victory even if there are differences in how to achieve it. The idea of protecting civilians to win their trust is a frequent theme recognized in many post–World War I counterinsurgency strategies and extends beyond Iraq and Afghanistan. Governments mostly shun the strategy of mass killings as they perceive that its costs outweigh the benefits. When they do pursue it, it is often because they are faced with information asymmetry that complicates correctly locating and punishing opponents.[123] World War I marked the era where the armies' direct contact with population was replaced with machine war, or manpower with vehicles such as tanks, aircraft, and trucks.

This change has "removed" the army from direct contact with the locals and made it difficult for the government to have informational advantages and be less selective in inflicting punishment.[124] Mass killings, many of which have resulted from this transformation, are by their very nature not selective in their punishment and generate the very costs that hurt the governments' chances against the insurgents. That many governments are aware of this is supported by a study of governments' tactics from 1945 to 2008 against insurgents. Although mass killings are not extinct as an approach, in 70% of all cases when guerrillas posed a major threat to the governments, the leaders chose not to pursue highly repressive tactics, which suggests that governments recognize the costs.[125]

From those who fought communist insurgents in the post-colonial period such as British general Frank Kitson (Malayan Emergency guerrilla war, 1948–1960) and French military leader David Galula (Algerian War, 1956–1958) to insurgent leaders such as Che Guevara, there has been a recognition among many leaders that winning popular support is a critical element of military effectiveness.[126] The fundamental idea of protecting the population to gain their support was recognized by President John F. Kennedy in 1961 and ultimately applied by US forces in Vietnam from 1965 to 1972 when Marines (Combined Action Platoon) provided security to local villagers.[127] In 2017, the Nigerian government invested in providing protection to farmers to allow them to safely cooperate with the government against Boko Haram insurgents.[128] In some contexts, governments start with a repressive strategy but alter parts of their approach when they discover its political limitations. While Turkey has been successful in reducing the strength of PKK in its three-decade struggle against the group, in 2018 and 2019 the government began experimenting with a hearts and minds approach in rural provinces by improving security in the area and raising expectations of economic investment in order to reap political support from the Kurds.[129]

There is variation among governments in how to handle civilians and how to merge offensive tactics with population safety, but many governments recognize that some aspect of restraint matters.[130] The difference is more about how to execute the strategy than in recognizing that basic effort needs to be made toward the population. For example, governments might ponder whether offering protection to civilians might suffice or whether a deeper engagement involving economic, social, and political support is necessary to win their support.

Although the debate about the best COIN approaches is likely to continue, in light of the evidence in favor of securing the support of the locals to help end wars through restraint in killings, we consider human rights protection to be a factor of military effectiveness that merits our attention. Given that security is a fundamental human need, our focus is on the most basic yet critical component of population-centric COIN that could make a difference in whether or not the locals join the insurgency.[131] We do not focus on other forms of legitimacy building such as economic and political reforms, which are more difficult to implement, widely debated as to their utility, and are mostly beyond the scope of PMSCs' involvement in civil wars. Consequently, we argue that PMSCs whose behavior in conflicts mirrors the expectations laid out in humanitarian law are more likely to be militarily effective than PMSCs with limited humanitarian standards, assuming other dimensions of military effectiveness we outlined here remain constant. It is important to recognize that some degree of flexibility is necessary in conflict zones and adherence to human rights standards should not be equated with complete pacifism or more extreme application of population-centric COIN. In Afghanistan, for example, the United States first focused almost exclusively on capturing the insurgents, with limited concern for local population. When the strategy backfired, a shift toward population-centric COIN followed. Yet the pendulum swung excessively in the other direction when commanders adopted very rigid rules of engagement to avoid civilian casualties only to see the loss of military gains. Changes were made and a more balanced approach was embraced under General Petraeus, who gave commanders more autonomy in responding to complex situations.[132]

Adherence to humanitarian law involves a commitment not to alienate the locals through excessive and systematic use of force and intimidation. Isolated episodes of aggressive behavior are unavoidable and may be necessary in some instances. In fact, the proportionality provision of the international humanitarian law focuses on excessive casualties relative to anticipated military advantage and therefore reflects the reality of war in which civilian casualties may occur. It is feasible that in some contexts even low-level human rights abuses, such as aggressive and arrogant behavior by PMSC employees, might be considered offensive to the locals and connected with spikes in insurgent support.[133] Yet it is difficult to isolate the independent effect that abuses on the lower end of the spectrum might pose for damaging popular support from more severe forms of violations; therefore our focus is on the higher threshold for abuses identified by international humanitarian law.

It must be noted that adherence to international humanitarian law is more likely to be effective when implemented systematically rather than incorporated as a measure of last resort. Consider, for example, the case of Iraq where PMSCs began to embrace stricter rules about aggressive behavior, yet the impact of this strategy on military effectiveness and subsequent termination of conflict has been questionable. After 2005, the number of civilian deaths at the hands of PMSCs has been on decline, in part due to somewhat greater oversight from the US military as evident by the establishment of the Reconstruction Operation Center and Six Operation Cells that prioritized common procedures and increased control of contractor movement[134] and in part due to media revelations of human rights abuses and the ensuing shareholder pressure to modify behavior. Unfortunately, greater commitment to human rights standards has not brought immediate changes to conflict dynamics. The failure, however, does not imply that commitment to humanitarian standards as a strategy that contributes to the termination of insurgencies is ineffective in shifting the tide of war in favor of the government. Rather, greater commitment to human rights protection from the private military and security industry came after the locals had already been alienated, a damage that cannot be easily offset. Gaining the trust of the locals requires a systematic approach from the early stages of war. Trust is a sensitive matter. It can be diminished easily and requires long-term investment to be recovered. Being exposed to civilian violence hardens attitudes against the out-group.[135] Humanitarian aid delivered to Pashtuns in Afghanistan as part of COIN proved less effective in regaining their support after out-group victimization and exposure to civilian violence had occurred.[136] In the case of El Salvador, the government's late implementation of population-centric COIN began to yield results in reducing the number of supporters that the insurgents were able to recruit and ultimately contributed to the termination of conflict because it was instituted by a newly elected civilian government that was perceived less as an out-group and was thus considered more credible.[137]

Another limitation of COIN concerns the issue of coordinating humanitarian efforts between government forces and PMSCs. Even if PMSCs emphasize commitment to upholding human rights, there is a possibility that their work could be undermined if the military is slow to adopt such practices and commits misdemeanors. In coordinated missions the locals might be unable to differentiate between the military and PMSC personnel, therefore criminal behavior by the military could be attributed to PMSCs and negatively affect the companies' performance in the long run. This is a possibility that could emerge in states where armies lack professionalism. However, it is less problematic when PMSCs assist troops from the West, as the latter are likely more tuned in to local sensitivities. In such contexts, the problem mostly lies with the contractors and not the military.[138]

Finally, the focus on PMSCs' interest in the population-centric approach, specifically on the level of restraint in the area of human rights abuses, depends not only on the companies' willingness to adhere to such standards but also on the extent to which the companies have a chance to make a difference in this area in the first place. It is possible that a client might disregard the benefits of COIN on some occasions and choose a more radical military option even if a PMSC suggests such an approach. For example, it would be difficult to imagine that Bashar al-Assad, who has targeted his population with chemical attacks and barrel bombs in the Syrian civil war,[139] would choose to work with a PMSC that pushes for population-centric COIN. In such cases the benefits of COIN as a strategy of military effectiveness would not be relevant. Such cases, however, are likely to be limited. Because international PMSCs recognize the value of working for mega consumers such as the UK and US, they would be reluctant to provide services to regimes that push for a drain-the-sea approach. Furthermore, PMSCs exert considerable influence over strategy design in many cases and thus are in a position to advocate COIN over other approaches to achieve military effectiveness. This is especially true in weak states, which rely on such companies to develop effective strategies for winning the war and on professional training of the military. Even if weak states are not specifically interested in a COIN approach, international PMSCs have an incentive to abide by international norms for tactical and financial benefits.[140] While the extent to which they push for it will vary depending on companies' corporate structure and local competition, overall, concern for reputation is likely to result in at least some emphasis on basic COIN principles. For example, in the 1990s Executive Outcomes supported winning over local population to gain their help with intelligence.[141]

In the context of strong states, here too, PMSCs have an opportunity to contribute to military effectiveness by honoring international humanitarian law through their influence on service design and the quality of service delivery. When analyzing contracts awarded to PMSCs in Iraq, Dickinson (2007) found vagueness and flexibility, which creates the very opening for armed contractors to influence aspects of their mission. Under the 2000 agreement between the US government and CACI, the procuring agency did not request specific tasks. Instead the agreement carried a general scope of "interrogation support and analysis for the US Army in Iraq."[142] While some companies may not automatically use the opening to advance more-humanitarian approaches in more-open contracts or the extent to which they do so may vary, such opportunity nevertheless exists. In cases where PMSCs' tasks are specifically outlined in the contract, PMSCs also have a chance to contribute to the success or failure of population-centric COIN in many areas. Although not all PMSC tasks are compatible with this aspect of COIN, troop training, security provision, intelligence, and interrogation all involve possible contacts with the locals. These are the tasks for which international PMSCs are often hired by governments in civil wars. It is also worth noting that most international PMSCs in our data set intervene to provide multiple services to the government and thus they are likely to be in charge of a task that requires interactions with the locals. When opportunity for these interactions exists, there is a strong likelihood that how a PMSC performs its task will contribute to the success of the mission.

What does it mean for PMSCs to develop a corporate commitment to humanitarianism that could enhance their military effectiveness? Such a commitment stems from a professional corporate culture that addresses the challenges of humanitarianism in the field and supports good practices in this area while recognizing the need to balance the goal of capturing an enemy with the needs of the locals. When the CEOs embrace such a culture, they signal to their employees that securing the support of the locals is as important as other aspects of military strategy that they are helping the government put in motion. A professional corporate culture will not diminish all instances of human rights abuses, but it will reduce systematic abuses when the company relies on threats and incentives and possibly on persuasion and normative change—although the latter may be harder to achieve—to modify employee behavior in conflict zones. Most of all, corporate commitment to humanitarianism entails a systematic approach toward the locals that, while flexible enough to accommodate the ever-evolving conflict dynamics, is not completely ad hoc in nature. As the cases of Iraq and Afghanistan illustrate, in the context of civilian deaths, past abuses could undermine the effectiveness of population-centric COIN even when greater accountability emerges later. In the subsequent chapter we examine why and how market pressure pushes some companies to cultivate a culture that reflects concerns for humanitarianism in conflict zones.

## Data on PMSCs' Interventions into Civil Wars

To examine how local competition and industry-wide competition that have pushed some companies to adopt a more transparent corporate structure affect PMSCs' military effectiveness in civil wars, we rely on our data of PMSCs' interventions into civil wars from 1990 to 2008. In cases when a civil war has been ongoing as of 1990, the start point for our data, we include the full duration of such a war. Thus the onset of some conflicts in our data set predates 1990. We define a *private military and security company* as a legally registered international corporate entity with a clear business structure that delivers military and security services for monetary compensation.[143] A PMSC is sometimes referred to in the literature as a private security company (PSC) or a private military company (PMC). Our focus is on any PMSC operating in a civil war that is (a) a legally registered international corporation; and (b) has a business structure involving some form of managerial hierarchy, for example, a CEO, vice president of operations, vice president of business development, or technical recruiter.

We focus on companies that support governmental troops in direct combat, deliver logistical support, provide security for airports and resource areas such as mining fields, train the military or police, and provide intelligence and communications services because each of these services can help tip the balance of power in a conflict. At the same time, we recognize that some services, such as providing more direct combat, could offer greater benefits for war termination than non-combat-related tasks and examine this point empirically in Chapter 5. PMSCs in our data set are international corporations. Although indigenous security providers have been proliferating on the market, they are likely to face different competitive pressures from international PMSCs and thus merit a unique theoretical approach. International PMSCs are more likely to operate in high-risk environments and provide lethal force–related services than are domestic PMSCs, many of which operate in a lower-risk environment. As result, the two types of actors are still likely to serve different markets. Furthermore, in some contexts domestic markets are monopolized by indigenous companies that are shielded from open competition with international PMSCs, as was the case in China prior to 2010.[144] In these types of markets, indigenous actors face different constraints than in more open competitive markets. A unique theoretical story is necessary to capture how this difference affects behavior. To illuminate the point about different market pressures, it is worth considering international initiatives such as the Montreux Document that have aimed to increase accountability of the private military and security industry. The Document sets expectations for PMSCs to respect international humanitarian law in armed conflict and was developed in response to the growing presence of transnational PMSCs. As such, it is possible that because such initiatives have been popularized in the past years, international PMSCs may be evaluated more on their adherence to international humanitarian law than indigenous companies when awarded contracts in open bidding.

We distinguish international PMSCs from other security providers, including mercenaries that sell their services on the military market without a corporate affiliation, ad hoc or temporary security providers, and armed support groups—such as those that have proliferated in Afghanistan—that are not legally registered corporate entities. For example, an Afghan company, Afghanistan Navin, that had contracts with Bagram Airbase and provided convoy escorts would not be considered a PMSC according to our definition, and therefore would not be included in our data set because it is a domestic company that is not registered.[145] It is important to note that PMSCs and other agents that deliver military and security services share some notable similarities, mainly in their profit-making goals and in demonstrating that the use of force can move from state control to the private sector. The current definition of mercenaries, as listed in Article 47 of the 1977 Protocol Additional to the 1949 Geneva Convention, considers the cumulative and concurrent requirements. Thus mercenaries are (1) recruited to fight; (2) take part in hostilities; (3) work for private gain; (4) are promised payment above that is given to members of armed services; (5) are not sent by any state that is party to the conflict; (6) are not part of the armed

forces; and, lastly, (7) are not nationals of a party to the conflict.[146] This definition excludes many PMSCs from being labeled mercenaries. The majority
of them do not engage directly in hostilities, although other characteristics
of mercenaries overlap with those of PMSCs. The legal definition would
not distinguish those PMSCs that have participated in hostilities, such as
Executive Outcomes, from mercenaries. In fact, the legal differentiation says
virtually nothing about the business organization of the two types of actors,
although business structure is recognized by many as an important difference between PMSCs and mercenaries. Modern-day PMSCs are more likely
than mercenaries to have a clear business structure that includes hierarchical
organization of management.[147] PMSCs pay taxes and can document at least
some of their activities.[148] Singer (2001–2002) notes the distinction between
PMSCs and such groups:

> The critical analytic factor is their modern business form. PMFs [private
> military firms] are hierarchically organized into incorporated and reg-
> istered businesses that trade and compete openly on the international
> market, link to outside financial holdings, recruit more proficiently than
> their predecessors, and provide a wider range of military services to a
> greater variety and number of clients.[149]

Because in many cases developing this form of business structure requires
upfront investment in time and resources, there is greater likelihood that
PMSCs have a long-term goal of surviving on the market and are more sensitive to market pressures emerging from growing competition. By contrast,
when groups operate on ad hoc basis without an organizational structure in
place or as a single, freelance mercenary, the cost of losing a service opportunity is likely to be lower than for PMSCs. In fact, O'Brien (2007)[150] argues
that operating with the goal of long-term survival is one of the key factors
that separate PMSCs from mercenaries. Former mercenaries who might become employed by a PMSC also face more pressure to deliver a reputable
service when part of the corporate entity than when working alone because
revelations of failures are increasingly more costly to PMSCs' reputations
and long-term survival. There is, of course, variation among PMSCs in their
corporate structure and responsiveness to market pressure, the argument
we advance in this book. Not every PMSC is likely to face the same level of
constraints in the marketplace. Yet overall, international PMSCs are more

likely to have a long-term perspective on survival, and this can make them more responsive to market pressures than other types of actors that provide security and military services.

In our data collection we build upon the work of Akcinaroglu and Radziszewski (2013),[151] who compiled data on PMSCs' interventions into African conflicts until 2008. We then used newspaper databases such as ProQuest, LexisNexis, scholarly articles, books, reports, and blogs to gather information on international PMSCs serving in conflicts in the rest of the world.[152] Gathering information on PMSCs' presence is challenging. While PMSCs advertise some of their activities on their websites, they often do not list the scope or duration of their contracts. To increase the reliability of our data, we occasionally consulted with country experts and NGOs to verify information about PMSCs' presence in specific civil wars. There were some instances where the contract was initially awarded to a PMSC only to be terminated immediately in its aftermath. Such was the case, for example, in Papua New Guinea in 1996 when the government contracted Sandline to train Papuan Defense Forces, assist with intelligence gathering, and conduct military operations against rebels in Bougainville. When the employees of Sandline were given higher command over their Papuan counterparts during field operations, the ensuing dissatisfaction among the military was the main reason behind the government's decision to cancel the contract.[153] In this conflict, Sandline was not given a real chance to perform the tasks, thus we chose not to code its presence in Papua New Guinea in 1996. All PMSCs in our data set have been given an opportunity to fulfill their contracts. To increase reliability, we relied on two individuals to collect data independently on the same data set.

While the use of PMSCs became prevalent in the post–Cold War era, there are also a few occasions in our data where we traced the existence of PMSCs to the 1980s in some conflict zones. Keenie Meenie Services (KMS), serving in Sri Lanka from 1984 to 1987, was one such case and the earliest example of PMSC presence in conflicts that have been ongoing as of 1990. This PMSC was hired to train the Special Task Force arm of the Sri Lankan military fighting with Liberation Tiger of Tamil Ealem.[154] Given that the conflict was ongoing as of 1990, the starting point of our data collection, we included any PMSC presence in such ongoing conflicts. Yet the rise in PMSC presence is civil wars is most notable in the post–Cold War environment. Pre-1990 PMSCs were often ad hoc groupings of former soldiers and these freelance mercenaries were not subject to institutional rules in the way that specialists working under the banner of a PMSC are. Before the end of the Cold War, recruitment of mercenaries was made on an individual basis and these hired men did not care for whom they worked. As such, the concept of accountability was not really applicable. Caring only about individual profits, freelance mercenaries were not disciplined nor were they accountable to their superiors the way employees in PMSCs are in a hierarchically organized structure. French mercenary Bob Denard and his notorious coup attempts in Africa exemplify such distinction. The overthrow of the Soilih government by Denard's mercenaries in 1978 demonstrates limited interest in public accountability. The freelance mercenaries also followed no rules and regulations pertaining to work ethics. While hired by the French government, the mercenaries eventually turned on their patron and started working for the South African government, which offered them more lucrative terms.[155] Thus, mercenaries who worked during the Cold War period and PMSCs that have emerged in the post–Cold War milieu are widely different entities. Our book focuses exclusively on PMSCs and not on freelance mercenaries.

Any analysis of PMSCs' impact on civil war termination is heavily reliant on definitions of war, its onset, and termination dates. We use two different data sets, Correlates of War (COW) Intra-State War Dataset v.4.0[156] and UCDP/PRIO Armed Conflict Dataset v.4.0,[157] to construct our observations on civil wars in the world. While the COW Intra-State data set uses a higher threshold of battle deaths, specifically 1,000, to code the onset and duration of civil wars, the UCDP/PRIO Armed Conflict Dataset includes minor-level conflicts, with a threshold of 25 battle deaths. The definition of civil war according to the COW Intra-State Dataset also includes effective resistance, the idea that the weaker side should be able to inflict at least 5% of the number of fatalities on its stronger opponent.

With the exception of these differences, both data sets define a *civil war* as a violent conflict internal to a country where the government is one of the fighting parties. We adopt this definition in our data. Nearly all of the civil wars in our data set are insurgencies, a type of civil war involving guerrilla tactics. There is a trend in the literature to use the UCDP/PRIO Armed Conflict Dataset in civil war studies in large part because many of the recent conflicts that have been of interest to scholars have failed to satisfy the criteria for battle deaths for major conflicts. Both Afghanistan and Iraq are classified as internationalized armed conflict, or an armed "conflict that occurs between the government of a state and one or more internal opposition group(s) with intervention from other states on one or both sides," in the UCDP/PRIO data,[158] starting in 2004 for Iraq and 2001 for Afghanistan. We include these conflicts in our minor conflict data set and code them as ongoing as of 2008. On the other hand, the conflicts listed in the COW Intra-State Dataset are the ones that come with heavy financial, human, and political costs and their termination is of utmost importance to domestic and international policymakers. Recent conflict in Afghanistan is classified as interstate war in the COW data for 2001, while the conflict in Iraq is listed as an interstate war in 2003; when the conflict is transformed into a civil war or intrastate conflict in the subsequent years, it is no longer listed in COW because the 1,000 battle-death threshold is not met for the civil war phase by 2008, the endpoint for our data.[159] Given that both Iraq and Afghanistan are classified as international conflicts and their transformation into a civil war does not meet the casualty threshold by 2008, these conflicts are not included in our data set for major wars.[160]

Analyzing our arguments in the context of minor and major civil wars offers specific advantages. Using these two data sets enables us to examine how the impact of local environment and corporate structure on PMSCs' accountability and subsequent impact on conflict duration differs in two types of conflicts. In major conflicts that are insurgencies both guerrilla warfare and conventional tactics are likely to be employed. Direct engagements may give PMSCs an opportunity to make substantive gains against the rebels. Such was the case when Executive Outcomes (EO) engaged in combat operations against the Revolutionary United Front (RUF) rebels in Sierra Leone, successfully repelling the rebels' incursion[161]. Eventually RUF had to sign the Lusaka Protocol in 1995 after a series of defeats. Often PMSCs do not directly engage in combat operations or assist the governments' troops during offenses, but are hired to provide intelligence, logistics or training services to restructure their client's armed forces, all of which are vital to the government's operations against rebel forces. In minor conflicts, however, guerrilla warfare is likely to be more prevalent than in major conflicts that are classified as insurgencies with much lower instances of conventional war. PMSCs have diversified and expanded their range of services to remain competitive and innovative in minor conflicts against guerrilla forces and terrorists[162]. Due to greater conflict complexity, it may take longer for PMSCs to help bring minor conflicts to an end compared to major conflicts even if the companies face market pressure that encourages greater accountability.

Our data include 73 conflicts in 39 countries based on the COW Intra-State War Dataset and 150 conflicts in 65 countries based on the UCDP/PRIO Armed Conflict Dataset. We do not count criminality, including drug-related violence, as civil wars. While organized crime and rebel groups may share some organizational and operational characteristics, they differ in many important respects. Rebel groups are political actors seeking political goals such as addressing grievances, regime destruction, or political inclusion. Definitions of civil wars often come with the inclusion of political objectives, whereas most criminal organizations lack explicit political objectives. One can challenge this notion by referring to the existence of criminal civil war groups, the so-called new civil wars that flourished in the aftermath of the Cold War. Schedler (2013),[163] for example, notes: "The new Mexican civil war is not a classical civil war in which ideological insurgencies strive to topple state power. It is a prototypical 'new' civil war, fought for material gain not social justice." However, the civil war as organized crime model has been critiqued extensively and convincingly in the civil war literature. First, empirical studies regarding criminal or predatory behavior lack robustness and suffer from measurement error, endogeneity problems, and clarity of causal mechanisms.[164] It is therefore extremely difficult to argue that when rebel organizations resort to crime, they have no political goals. For example, ideology has been in the forefront even in the FARC insurgency, which has benefited extensively from the exploitation of cocoa production for financing its war.[165] Furthermore, unlike organized crime, rebel movements often challenge the status quo rather than work within it. Criminal groups are mainly interested in preserving the political status quo and co-opting existing political institutions rather than subverting them.[166] In light of the different goals of the main actors, we consider criminal violence as a different phenomenon from civil wars involving rebel movements.

The final issue to address is the coding of the duration of each conflict, a contentious issue in the civil war literature.[167] Decisive outcomes such as victory as well as peace agreements create clear-cut endpoints for the termination of conflict, yet military victories and peace treaties are relatively rare in civil wars.[168] Oftentimes, the number of fatalities fails to satisfy the defined threshold of the data set and hence on the outlook, it seems as if the conflict is terminated even if the country is not at peace. One of the challenges is to decide whether a rebel group whose activities oscillate between low and high activity signals the continuation of conflict or its end. Coding the conflict as "terminated" each year the number of battle deaths fails to satisfy the defined threshold in the data set creates too many onsets of conflict. This is specifically true for minor conflicts. For that reason, we adopt the commonly used criteria in the literature and code such cases as "termination of conflict" if the rebel group is inactive, unable to satisfy the minimum threshold of battle deaths, for at least two successive years.[169]

Tables 2.1 and 2.2 show the major and minor civil wars that are included in our data set along with PMSCs' interventions in each conflict. We list the

**Table 2.1** PMSCs' Interventions into Major Conflicts (1,000 battle deaths)

| Warring Groups | War Onset/ End | PMSC Intervention(s) | Year |
|---|---|---|---|
| Afghanistan/, Mujahaden | 1989–2001 | None | |
| Algeria/Islamic Front, AIS | 1992–1999 | Eric SA | 1992 |
| Angola/UNITA | 1976–1994 1998–2002 | Executive Outcomes Capricorn Teleservices Omega Support Stabilico Panasec Corporate Dynamics International Defense and    Security Ltd (IDAS) IRIS Service AirScan LR-Avionics | 1992–1994 1994 1994 1998 1998 1998 1998 1998 1998 2001 |
| Azerbaijan/ Nagorno-Karabakh | 1991–1993 | None | |
| Bosnia/Bosnia Serb Rebellion | 1992–1994 | None | |
| Burundi/Tutsi Army | 1993–1998 | None | |
| Burundi/FNL, Frolina | 2001–2003 | None | |
| Cambodia/Khmer Rouge | 1989–1991, 1993–1997 | None | |
| Chad/Deby's MPS | 1989–1990 | None | |
| Chad/MDD, MDJT | 1998–2000 | None | |
| Chad/FUDC | 2005–2006 | None | |
| Congo Brazzaville/ FDU(Cobra militias) | 1997 | Levdan AirScan | 1997 1997 |
| Congo Brazzaville/ Ninjas and Cocoye militias | 1998–1999 | None | |
| Cote d'Ivoire/ MPCI, MPIGO & MJO | 2002–2004 | None | |
| Croatia/Krajinia Serbs | 1995 | MPRI | 1995 |

**Table 2.1** *Continued*

| Warring Groups | War Onset/ End | PMSC Intervention(s) | Year |
|---|---|---|---|
| Democratic Republic of the Congo/AFDL | 1996–1997 | Omega Support | 1996–1997 |
| | | MPRI | 1996–1997 |
| | | Kellogg Brown and Root | 1996–1997 |
| | | Geolink | 1997 |
| | | Executive Outcome/Sandline | 1997 |
| | | Stabilico | 1997 |
| | | Intercon | 1997 |
| | | International Defense and Security (IDAS) | 1997 |
| Democratic Republic of the Congo/RCD & MLC et al. | 1998–2002 | Defense Systems Ltd. (DSL) | 1998 |
| | | Safenet | 1998 |
| | | IRIS Service | 1998 |
| | | Executive Outcomes (spinoffs) | 1998 |
| El Salvador/ Salvadorian Democratic Front | 1979–1992 | None | |
| Ethiopia/ Eritrea | 1982–1991 | None | |
| Ethiopia/ OLF | 1999 | None | |
| Georgia/Reform | 1991–1992 | None | |
| Georgia/Abkhazia | 1993–1994 | None | |
| Guinea/RDFG | 2000–2001 | None | |
| India/Kashmir Insurgents | 1990–2005 | None | |
| Indonesia/GAM | 1989–2003 | None | |
| Iraq/Shiites and Kurds | 1991 | None | |
| Iraq/ PUK | 1996 | None | |
| Lebanon/militias | 1989–1990 | None | |
| Liberia/NPLF | 1989–1990, 1996 | None | |
| Liberia/NPLF, Ulimo | 1992–1995 | MPRI | 1995 |
| Liberia/LURD & MODEL | 2002–2003 | None | |
| Moldova/Dniestra | 1991–1992 | None | |
| Mozambique/ Renamo | 1979–1992 | None | |

*Continued*

**Table 2.1** *Continued*

| Warring Groups | War Onset/ End | PMSC Intervention(s) | Year |
|---|---|---|---|
| Nepal/CPN | 2001–2006 | None | |
| Nicaragua/ Contras | 1982–1990 | Enterprise Eagle Aviation Services and Technology, Inc. EAST/Saladin | 1985–1986 |
| Pakistan/Waziri | 2004–2006 | DynCorp | 2005–2006 |
| Papua New Guinea/ Bougainville Secession | 1989–1992 | Defense Systems Limited (DSL) -terminated | 1993 |
| Peru/Shining Path | 1982–1992 | None | |
| Philippines/NPA | 1972–1992 | None | |
| Philippines/MILF | 2000–2001 | None | |
| Philippines/ MILF,ASG | 2003 | Control Risks Group | 2003 |
| Philippines/MILF, NPA | 2005–2006 | Control Risks Group Grayworks | 2005–2006 2005 |
| Russia/Chechnya | 1994–2003 | None | |
| Rwanda/RPF | 1994 | Ronco | 1994 |
| Rwanda/Hutu Rebels | 1997–1998 | None | |
| Rwanda/Army for the Liberation of Rwanda | 2001 | None | |
| Sierra Leone/ Kabbah faction | 1998–1999 | Executive Outcomes Sandline Lifeguard Management Cape International Corporation ICI (Pacific Architects Engineers) | 1998 1998 1998 1998 1998 |
| Sierra Leone/RUF | 1991–1996 | Specialist Services International Marine Protection Executive Outcomes Ibis Air Gurkha Control Risks Group Group 4 Defense Systems Limited (DSL) Sandline Lifeguard Management Teleservices | 1991 1992 1995–1996 1995–1996 1995 1995 1995 1995 1996 1996 1996 |

**Table 2.1** *Continued*

| Warring Groups | War Onset/ End | PMSC Intervention(s) | Year |
|---|---|---|---|
| Somalia/rebel clans | 1988–1991 | None | |
| Somalia/Aideed Faction, USC | 1992–1997 | None | |
| Somalia/SCIC | 2006–2008 | ATS Tactical and Select Armor Secopex | 2006–2008 2008 |
| Sri Lanka-Tamils | 1983–2002 | Keeni Meeni Services Maritime Trident | 1984–1987 2000 |
| Sudan/SPLA-Garang Faction | 1983–1991 | None | |
| Sudan/SLA, JEM | 2003–2006 | None | |
| Tajikistan/ United Tajik Opposition | 1992–1997 | None | |
| Turkey/ Kurds | 1991–1999 | None | |
| Yemen/South Yemen | 1994 | None | |
| Yemen/Zaidi Muslims | 2004, 2007 | None | |
| Yugoslavia/ Croatian Independence | 1991–1992 | None | |
| Yugoslavia/KLA | 1998–1999 | MPRI[a] | 1999 |

*Note*: a-alleged

warring parties and the dates for war onset and termination. In cases when the government fights the same group over the course of the years, we consider such conflict to be the same unless two years of inactivity have passed, in which case, a new war begins.[170]

The frequency of interventions has been on the rise since 1990. Figure 2.2 shows the dramatic spike in PMSCs' presence in major civil wars in 1998, when 15 private companies delivered services in conflict zones. After 1998, PMSCs' interventions have declined in both major wars and minor wars but the decline has been brief for minor wars, instead rising with 2003 (Figure 2.3). This decline can be attributed to the lower frequency of major civil wars post 1998. As such, the demand for PMSC services in major wars was slowing down not because of developments associated with the industry itself but

**Table 2.2**  PMSCs' Interventions into Minor Conflicts (25 battle deaths)

| Warring Groups | War Onset/End | PMSC Intervention(s) | Year |
|---|---|---|---|
| Afghanistan/, former warlords, Mujahaden | 1978–2001 | None | |
| Afghanistan/ Taleban,Hizb-i Islami- yi Afghanistan* | 2003–2008 | Ronco Consulting | 2004–2008 |
| | | DynCorp | 2004–2008 |
| | | Gurkha | 2004–2008 |
| | | Kellogg Brown and Root | 2004–2008 |
| | | Blackwater | 2004–2008 |
| | | Lockheed Martin | 2004–2008 |
| | | Armour Group | 2004–2008 |
| | | Hart Security | 2004–2008 |
| Algeria/Takfir wa'l Hijra, AIS, GIA, AQIM | 1991–2008 | Eric SA | 1992 |
| Angola/UNITA, FNLA | 1975–2002 | Executive Outcomes | 1993–1994 |
| | | Capricorn | 1994–1996 |
| FLEC, FLEC-FAC, FLEC/R | 1991–2008 | Teleservices | 1994–1996 |
| | | Alpha 5 | 1995–1996 |
| | | Saracen | 1996 |
| | | Omega Support | 1996, 1998 |
| | | Stabilico | 1996, 1998 |
| | | Panasec Corporate Dynamics | 1996–1998 |
| | | Bridge Resources | 1996 |
| | | Coin Security | 1996 |
| | | Corporate Trading International | 1996 |
| | | Shibata Security | 1996 |
| | | Longbeach | 1996 |
| | | Defense and Security Ltd. (DSL) | 1996 |
| | | MPRI | 1996 |
| | | International Defense and Security (IDAS) | 1997–1998 |
| | | AirScan | 1997–1998, 2002 |
| | | Iris Service | 1998 |
| | | Allerta | 1998 |
| | | LR-Avionics | 2001 |

**Table 2.2** *Continued*

| Warring Groups | War Onset/End | PMSC Intervention(s) | Year |
|---|---|---|---|
| Azerbaijan/ Nagorno-Karabakh | 1991–1994 | None | |
| | 2005 | Blackwater | 2005 |
| Azerbaijan/Opon forces | 1994–1995 | None | |
| Bangladesh/JSS-SB | 1975–1992 | None | |
| Bosnia/ Autonomous Province of Western Bosnia | 1993–1995 | None | |
| Bosnia/Bosnia Serb Rebellion | 1992–1995 | None | |
| Bosnia/Croatian irregulars | 1993–1994 | None | |
| Burundi/CNDD, Palipehutu, CNDD-FDD, Palipehutu-FNL, Frolina | 1991–2008 | None | |
| Cambodia/Khmer Rouge, Funcinpec, KPNLF, KNUFNS | 1978–1998 | Copras | 1993 |
| Chad/FAN, FAP, FAT, GUNT, CDR, Islamic Legion, MPS, Mosanat, CNR, CSNPD, FNT, MDD, FARF, MDJT, FUCD, RAFD, UFDD, AN | 1976–2008 | None | |
| Central African Republic/Forces of Francois Bozize | 2002 | None | |
| Central African Republic/UFDR | 2006 | None | |
| Colombia/FARC, ELN, M-19, EPL | 1964–2008 | Spearhead | 1988–1990 |
| | | DynCorp | 1993–2008 |
| | | Blackwater | 2005 |
| | | AirScan | 1997–2001 |
| | | MPRI | 2000–2001 |
| | | Lockheed Martin | 2002–2008 |
| | | ADC | 2001 |
| | | ARINC Inc. | 2002–2008 |
| | | TRW | 2002 |
| | | Matcom | 2002 |

*Continued*

Table 2.2 *Continued*

| Warring Groups | War Onset/End | PMSC Intervention(s) | Year |
| --- | --- | --- | --- |
| | | Cambridge Communications | 2002 |
| | | Virginia Electronic Systems | 2002 |
| | | AirPark and Sales Service | 2002 |
| | | Integrated Aero Systems | 2002 |
| | | Northrop Grumman | 2002–2008 |
| | | Alion LLC | 2002 |
| | | The Rendon Group | 2002 |
| | | ACS Defense | 2002 |
| | | INS | 2002 |
| | | Science Applications Intern. Corp (SAIC) | 2002–2008 |
| | | ManTech, ManTech International | 2002–2008 |
| | | Olgoonik | 2003–2008 |
| | | Oakley Networks | 2006–2007 |
| | | CCE-Construction, Consulting & Engineering | 2003–2007 |
| | | ITT | 2006–2008 |
| | | OPTEC | 2003–2007 |
| | | Telford Aviation | 2003–2008 |
| | | King Aerospace | 2003–2007 |
| | | CACI International | 2003–2008 |
| | | Tate Incorporated | 2003–2007 |
| | | Chenega Federal Systems | 2003–2007 |
| | | PAE | 2003–2008 |
| | | Omnitempus | 2006–2007 |
| | | DRS TAMSCO | 2008 |
| | | J&J Maintenance Colombia | 2008 |
| | | Raytheon Technical Service Co. | 2008 |
| Comoros/MPA- Republic of Anjouan | 1997 | None | |
| Congo Brazzaville/ Cobras, Ninjas | 1993–1994 | Levdan | 1994 |
| Congo Brazzaville/ Cobras, Ninjas and Cocoye militias, Ntsiloulous | 1997–2002 | Levdan | 1997 |
| | | AirScan | 1997 |

**Table 2.2** *Continued*

| Warring Groups | War Onset/End | PMSC Intervention(s) | Year |
|---|---|---|---|
| Cote d'Ivoire/MPCI, MPIGO & MJP, FN | 2002–2004 | None | |
| Croatia/Krajinia Serbs | 1992–1995 | MPRI | 1994–1995 |
| Democratic Republic of the Congo/RCD & MLC, RCD-ML, AFDL | 1996–2001 | Omega Support | 1996–1997 |
| | | MPRI | 1996–1997 |
| | | Kellogg Brown and Root | 1996–1997 |
| | | Geolink | 1997 |
| | | Executive Outcome/ Sandline | 1997 |
| | | Stabilico | 1997 |
| | | Intercon | 1997 |
| | | International Defense and Security (IDAS) | 1997 |
| | | Defense Systems Limited (DSL) | 1998 |
| | | Safenet | 1998 |
| | | IRIS Service | 1998 |
| | | Executive Outcomes (spinoffs) | 1998 |
| Democratic Republic of the Congo/CNDP | 2006–2008 | None | |
| Democratic Republic of the Congo/BDK | 2007–2008 | None | |
| Djibouti/FRUD | 1991–1994 | None | |
| Djibouti/FRUD-AD | 1999 | None | |
| Egypt/al-Gama'a al-Islamiyya | 1993–1998 | None | |
| El Salvador/ERP, EPL, FMLN | 1979–1991 | None | |
| Eritrea/EIJM – AS | 1997–1999 | None | |
| | 2003 | None | |
| Ethiopia/EPLF, ELF, ELF-PLF | 1964–1991 | None | |
| Ethiopia/ TPLF, EPDM, EDU, EPRP, EPRDF | 1976–1991 | None | |
| Ethiopia/OLF | 1977–2008 | None | |
| Ethiopia/ALF | 1989–1991 | None | |
| Ethiopia/ONLF | 1994–2008 | None | |
| Ethiopia/Al-Ittihad | 1995–1999 | None | |

*Continued*

**Table 2.2** *Continued*

| Warring Groups | War Onset/End | PMSC Intervention(s) | Year |
|---|---|---|---|
| Ethiopia/ARDUF | 1996 | None | |
| Georgia/Zvadists | 1991–1993 | None | |
| Georgia/Republic of Ossetia | 1992, 2004, 2008 | None | |
| | | Cubic Defense Applications | 2004 |
| | | MPRI, American Systems Corporation | 2008 |
| Georgia/Abkhazia | 1992–1993 | None | |
| Guatemala/FAR I, FAR II, ORPA, EGP, URNG | 1963–1995 | None | |
| Guinea/RDFG | 2000–2001 | None | |
| Haiti/FLRN, OP Lavalas (Chimères) | 2004 | None | |
| India/Sikh Insurgents | 1983–1993 | None | |
| India/Kashmir | 1989–2008 | None | |
| India/ABSU, NDFB | 1989–2004 | None | |
| India/PWG, MCC, CPI-Maoist | 1990–2008 | None | |
| India/ULFA | 1990–2008 | None | |
| India/NSCN-IM | 1992–2000 | None | |
| India/ ATTF, NLFT | 1992–2006 | None | |
| India/PLA, UNLF, KCP, PREPAK | 1992–2008 | None | |
| India/KNF | 1997 | None | |
| India/NSCN-K | 2005–2007 | None | |
| India/PULF | 2008 | None | |
| India/DHD-BW | 2008 | None | |
| Indonesia/ Fretilin | 1975–1992, 1997–1998 | None | |
| Indonesia/GAM | 1990–2005 | Bravo Security | 2002 |
| Iraq/ PUK, KDP , KDP-QM | 1973–1996 | None | |
| Iraq/SCIRI | 1991–1996 | None | |
| Iraq/Al Mahdi, Ansar al-Islam, ISI* | 2004–2008 | Aegis Defense | 2004–2008 |
| | | Agility Logistics | 2004–2008 |
| | | AKE Group | 2004–2008 |
| | | Allied International | 2004–2008 |
| | | Alfagates | 2004–2008 |

**Table 2.2** *Continued*

| Warring Groups | War Onset/End | PMSC Intervention(s) | Year |
|---|---|---|---|
| | | Ardan Consulting | 2004–2008 |
| | | Armour Group | 2004–2008 |
| | | Babylon Gates | 2004–2008 |
| | | Bearing Point | 2004–2008 |
| | | BH Defense | 2004–2008 |
| | | Black Hawks | 2004–2008 |
| | | Blackwater | 2004–2008 |
| | | BLP | 2004–2008 |
| | | Blue Hackle | 2004–2008 |
| | | Britam Defense | 2004–2008 |
| | | CACI International | 2004–2008 |
| | | Castleforce | 2004–2008 |
| | | Castlegate | 2004–2008 |
| | | Centurion Risks | 2004–2008 |
| | | Conchise Consultancy | 2004–2008 |
| | | CSC | 2004–2008 |
| | | Control Risks Group | 2004–2008 |
| | | Custer Battles | 2004–2008 |
| | | Dilligence LLC | 2004–2008 |
| | | Dimensions Int. | 2004–2008 |
| | | DTS Security | 2004–2008 |
| | | DynCorp | 2004–2008 |
| | | ECCI | 2004–2008 |
| Iran/KDPI | 1979–1996 | None | |
| Iran/MEK | 1986–1993 | None | |
| | 1997–2001 | None | |
| Iran/Jondullah, PJAK | 2005–2008 | None | |
| Israel, Palestinian insurgents, Non PLO groups, PLO, Rejectionist Front, Fatah, PFLP, PFLP–GC, PIJ, PNA, Hamas | 1949–1996 | None | |
| Israel/Fatah, PNA, Hamas, AMB, PIJ | 2000–2008 | DynCorp | 2008 |
| Israel/Hezbollah | 1990–1999, 2006 | None | |
| Laos/LRM | 1989–1990 | None | |
| Lebanon/LNM, Amal, NUF, Lebanese Army (Aoun), Lebanese Forces-Hobeika Faction | 1982–1990 | None | |

*Continued*

**Table 2.2** *Continued*

| Warring Groups | War Onset/End | PMSC Intervention(s) | Year |
|---|---|---|---|
| Liberia/NPFL, INPFL | 1989–1990 | | |
| Liberia/LURD & MODEL | 2000–2003 | None | |
| Macedonia/UCK | 2001 | MPRI | 2001 |
| | | AirScan | 2001 |
| Mali/MPA | 1990 | None | |
| Mali/FIAA | 1994 | None | |
| Mali/ATMNC | 2007–2008 | None | |
| Mexico/EZLN | 1994 | None | |
| Moldova/Dniestra | 1992 | None | |
| Mozambique/Renamo | 1977–1992 | None | |
| Myanmar/KNUP, KNU | 1949–2008 | None | |
| Myanmar/CPB, CPB–RF, PVO—"White Band" faction, ABSDF | 1948–1994 | None | |
| Myanmar/KIO | 1961–1992 | None | |
| Myanmar/NMPSP | 1990 | None | |
| Myanmar/ARIF, RSO | 1991–1994 | None | |
| Myanmar/KNPP | 1992, 1996, 2005 | None | |
| Myanmar/MTA, SSA-S | 1993–2008 | None | |
| Myanmar/BMA | 1996 | None | |
| Nepal/CPN | 1996–2006 | None | |
| Nicaragua/Contras | 1982–1990 | Enterprise Eagle Aviation Services and Technology, Inc. EAST/Saladin | 1985–1986 |
| Niger/FLAA | 1991–1992 | None | |
| Niger/CRA | 1994 | None | |
| Niger/FDR | 1995 | None | |
| Niger/UFRA | 1997 | None | |
| Niger/MNJ | 2007–2008 | None | |
| Nigeria/Ahlul Sunnah Jamaa | 2004 | MPRI | 2004 |
| Nigeria/NDPVF | 2004 | MPRI | 2004 |
| Papua New Guinea/ Bougainville Secession | 1989–1996 | Defense Systems Limited (DSL)-terminated | 1993 |
| Pakistan/MQM | 1990 | None | |
| | 1995–1996 | None | |

**Table 2.2** *Continued*

| Warring Groups | War Onset/End | PMSC Intervention(s) | Year |
|---|---|---|---|
| Pakistan/Baluchi insurgents | 2004–2008 | DynCorp | 2005–2008 |
| | | Blackwater | 2007–2008 |
| | | Catalyst Services | 2008 |
| Pakistan/TTP, TNSM | 2007–2008 | Blackwater | 2007–2008 |
| | | DynCorp | 2007–2008 |
| | | Catalyst Services | 2008 |
| Peru/Shining Path, MRTA | 1982–1999 | DynCorp | 1992–2008 |
| Peru/Shining Path | 2007–2008 | DynCorp | 2007–2008 |
| Philippines/CPP | 1969–2008 | None | |
| Philippines/MIM, MILF, MNLF, MNLF-NM, MNLF-HM, ASG | 1970–2008 | Grayworks | 2000, 2005 |
| | | Control Risks Group | 2003, 2006 |
| | | Blackwater | 2006 |
| Russia/ Republic of Armenia | 1990–1991 | None | |
| Russia/APF | 1990 | None | |
| Russia/ Parliamentary forces | 1993 | None | |
| Russia/Chechnya | 1994–2007 | None | |
| Russia/ Wahhabi movement of the Buinaksk district | 1999 | None | |
| Russia/ Caucasus Emirate | 2007–2008 | None | |
| Rwanda/RPF | 1990–1994 | Ronco | 1994 |
| Rwanda/FDLR | 1997–2002 | None | |
| Senegal/MFDC | 1990–2003 | None | |
| Sierra Leone/RUF, AFRC, Kamajors, WSB | 1991–2000 | Specialist Services Int. | 1991 |
| | | Marine Protection | 1992 |
| | | Executive Outcomes | 1995–96,98 |
| | | Ibis Air International | 1995–1996 |
| | | Gurkha Security Guards Ltd. | 1995 |
| | | Control Risks Group | 1995 |
| | | Group 4D | 1995 |
| | | ICI | 1997–1998 |
| | | Cape International | 1997–1998 |
| | | Sandline | 1997–1998 |
| | | Lifeguard Management | 1996–1998 |
| | | Teleservices | 1996–1997 |

*Continued*

**Table 2.2** *Continued*

| Warring Groups | War Onset/End | PMSC Intervention(s) | Year |
|---|---|---|---|
| Somalia/SNM, SPM, SSDF, UCS,UCS/SNA | 1986–1996 | None | |
| Somalia/SRRC | 2001–2002 | None | |
| Somalia/ARS/UIC, Al-Shabaab | 2006–2008 | ATS Tactical | 2006–2008 |
| | | Select Armor | 2006–2008 |
| | | Secopex | 2008 |
| Spain/ETA | 1991–1992 | None | |
| Sri Lanka-Tamils | 1984–2008 | Keeni Meeni Services | 1984–1987 |
| | | Maritime Trident | 2001 |
| Sri Lanka/JVP | 1989–1990 | None | |
| Sudan/SPLM/A, SAF, NDA, SPLA-Garang Faction, NDA, JEM, NRF, SLM/A, SLM/A - Unity, SLM/A-MM | 1983–2008 | AirScan | 1997–1998 |
| | | DynCorp | 2001–2002 |
| Tajikistan/United Tajik Opposition, Movement for Peace in Tajikistan | 1992–1996 | None | |
| | 1998 | None | |
| Thailand/Patani Insurgents | 2003–2008 | None | |
| Trinidad and Tobago/ Jamaat al-Muslimeen | 1990 | None | |
| Turkey/PKK | 1984–2008 | None | |
| Turkey/ Devrimci Sol | 1991–1992 | None | |
| Turkey/MKP | 2005 | None | |
| Uganda/Fronasa, Kikosi Maalum, UNLA, FUNA, NRA, UNRF, UFM, HSM, UPDA, Lord's Army, LRA, UPA, ADF, WNBF, UNRF II | 1979–2008 | EO | 1995 |
| | | Saracen | 1996–2008 |
| | | AirScan | 1997–1998 |
| | | MPRI | 2007–2008 |
| | | DynCorp | 2008 |
| United Kingdom/ PIRA | 1971–1991 | None | |
| United Kingdom/ PIRA | 1998 | None | |
| United States/ Al-Qaida | 2001–2008 | Blackwater | 2004–2006 |
| Uzbekistan/IMU | 1999–2000 | None | |
| Uzbekistan/JIG | 2004 | None | |

**Table 2.2** *Continued*

| Warring Groups | War Onset/End | PMSC Intervention(s) | Year |
|---|---|---|---|
| Yugoslavia/Republic of Slovenia | 1991 | None | |
| Yugoslavia/Croatian Independence | 1991 | None | |
| Yugoslavia/UCK | 1998–1999 | MPRI[a] | 1999 |
| Yemen/South Yemen | 1994 | None | |

*over 100 companies present in Iraq and Afghanistan, list is a sample to show the existance of multiple providers
[a] alleged

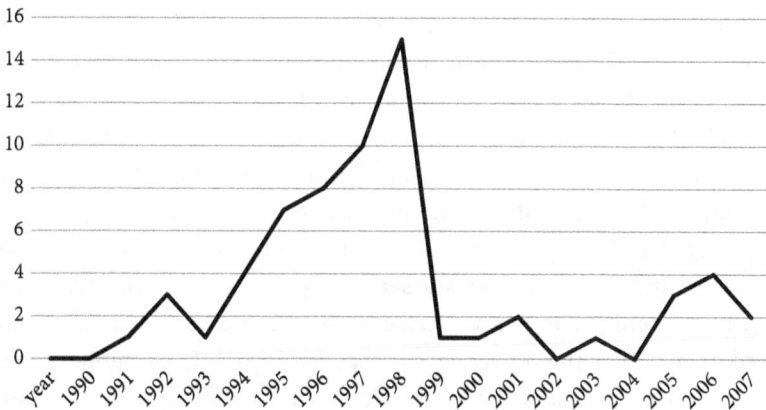

**Figure 2.2** Number of PMSCs in Major Conflicts

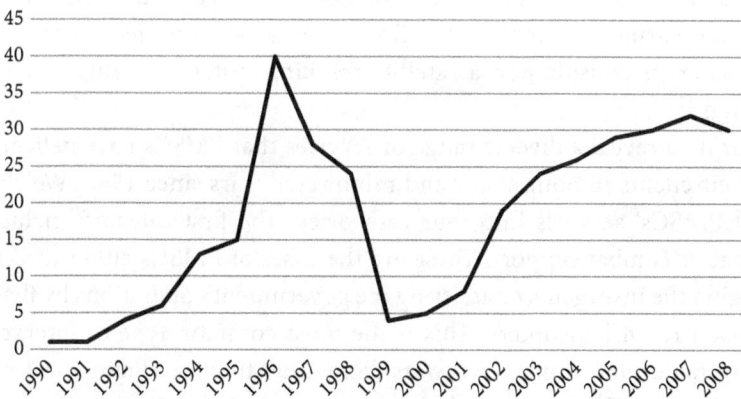

Note: We used an approximate number of PMSCs in Iraq and Afghanistan

**Figure 2.3** Number of PMSCs in Minor Conflicts

*Note*: We used an approximate number of PMSCs in Iraq and Afghanistan because of the decline in major conflicts from 1998–2008. Hence, there is a shift toward greater occurrence of PMSC interventions into minor conflicts after 1998, contexts in which guerrilla warfare dominates.

PMSCs deliver different types of services to fulfill their contracts, including direct combat, combat support, logistics, communications, technical support, intelligence, training, and security. The range of services that PMSCs deliver varies from mundane jobs such as cooking and cleaning to sophisticated tasks including maintaining weapons systems, air surveillance, providing combat pilots, securing airports, and counterterrorism training. This has led some authors to attempt to create a widely accepted typology in services. While some authors have classified PMSCs according to service type, however, there is no single categorization that is accepted by all. Singer (2003)[171] was the first to create a typology of PMSCs according to service provision by classifying them as military providers, military consultants, or military support firms. Avant (2005),[172] however, divided PMSCs into two groups based on their contract types, those that provided external military support and internal police support. Her first category included armed and unarmed operational support, logistics support, and military training, while the latter category included armed and unarmed site security, police training, crime prevention, and intelligence. Kinsey (2006) differentiates between companies based on whether the object that is secured is private or public and the means, range of lethality, used to secure it.[173] Leaving aside these discrepancies in typology, what makes our attempts to differentiate between firms especially difficult is the increasing trend in diversification of services. Control Risks Group (CRG), for example, expanded its consulting to include armed security. Olive Group has added a variety of services such as crisis consultancy, a satellite tracking system, intelligence, and training.[174]

Our data reveal a diverse range of services that PMSCs have delivered for their clients in both major and minor civil wars since 1990. We classified PMSCs' services into four categories. The first category includes combat or combat support. These are the cases of PMSCs either directly engaging the insurgents or assisting the government's operations by flying combat aircraft/helicopters. This is the most common type of intervention in major and minor civil wars from 1990 to 2008 (Figures 2.4 and 2.5). The second category includes security provision. In these types of

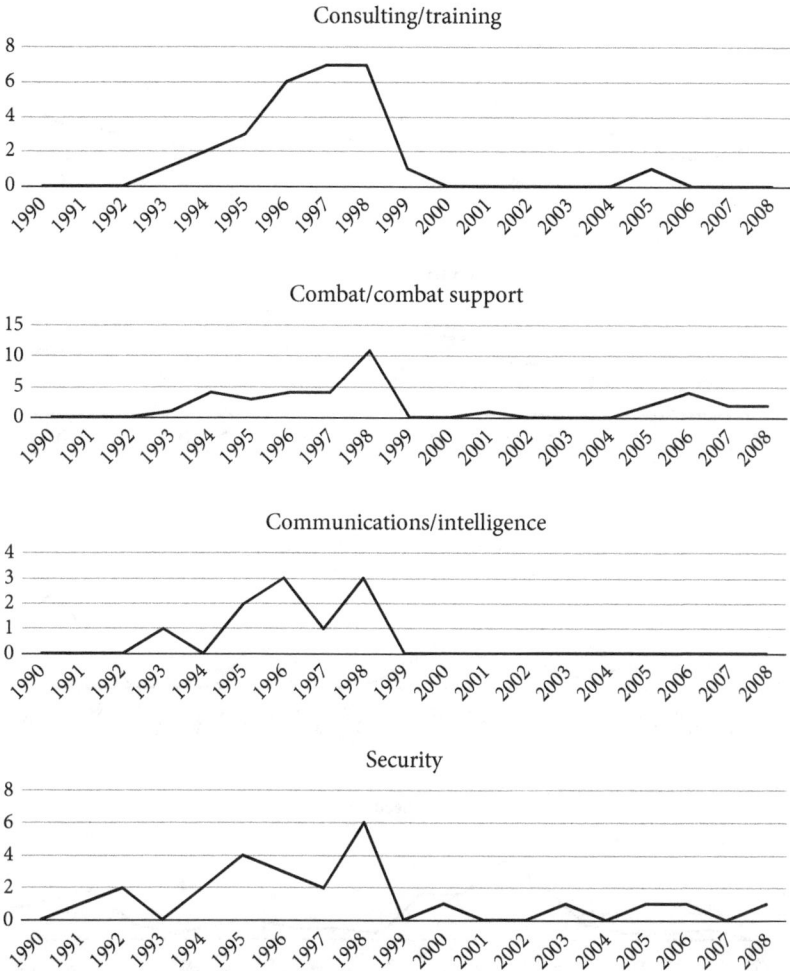

**Figure 2.4** Type of PMSCs' interventions into major conflicts: trends in service provision

*Note*: Iraq and Afghanistan are excluded, as the data we have on services in the two countries are limited.

interventions PMSCs provide security to government officials and/or protect valuable resource areas. This is specifically a frequent form of intervention in minor conflicts. The third type of service delivery involves consultancy and/or police/military training, while the fourth type of service falls into the category of communications and intelligence analysis,

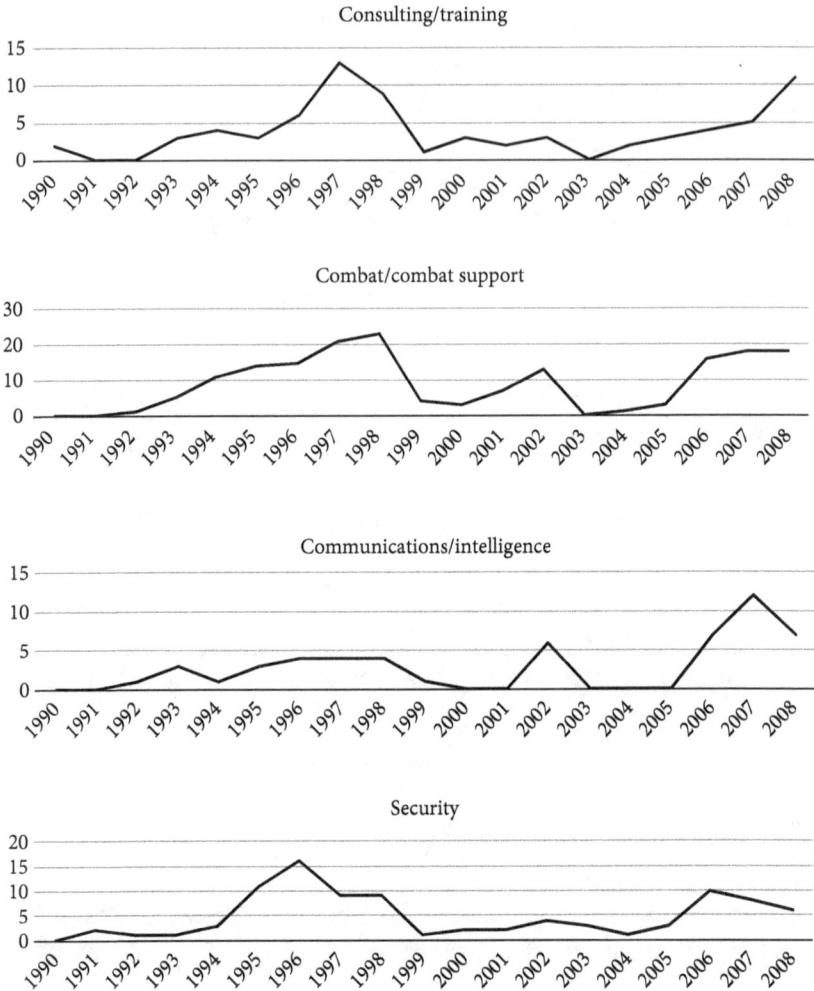

**Figure 2.5** Type of PMSCs' interventions into minor conflicts: trends in service provision

*Note*: Iraq and Afghanistan are excluded, as the data we have on services in the two countries are limited.

cases in which PMSCs deliver improved communications technology and/or help the governments develop a strategy in conflict. The demand for military and police training was considerably greater in major civil wars than in minor conflicts (62% of intervention years in major civil wars vs. 49% of intervention years in minor civil wars), possibly because such wars have involved more direct engagement between the military and rebel forces, thereby necessitating greater investment in military training than might be needed in minor conflicts.

## Conclusion

With the rise of the private military and security industry, questions have emerged about the possible consequences of the trend to outsource security to corporate actors. Existing debate has focused extensively either on highlighting negative aspects of the industry's growing presence in conflicts around the world or pointing to new and effective ways in which such actors could impact security around the world. What has emerged from this debate is an agreement that monitoring mechanisms are vital for improving PMSCs' performance overall. Emergence of international norms on hiring practices and expectations for contractors' behavior in armed conflict, ANSI/ASIS or ISO certifications, domestic regulations coupled with improvements in contract clarity, and market pressure that has pushed companies to self-regulate through the International Code of Conduct Association (ICoCA) and trade organization memberships have played a role in increasing accountability though not eliminating the problem. Limited focus on systematically analyzing PMSCs' performance in different types of civil wars has made it difficult to propose relevant policy recommendations about ways to improve corporate actors' accountability that reflect empirical findings across time and space. Compounding the difficulty is the absence of collaboration among researchers investigating private military and security industry and scholars of civil wars, especially those engaged in empirical research on civil war dynamics. The latter group has focused mostly on the impact of states and international or regional organizations on civil war duration,[175] with somewhat less attention devoted to the role of non-state actors, such as PMSCs, in terminating internal wars.

One of the goals of this book is to bridge the gap between research on PMSCs and empirical work on civil war dynamics by advancing a more systematic approach to analyzing conditions under which PMSCs could shape the duration and termination of wars. By focusing on conditions under which market dynamics, specifically variation in local competition and the growth in global competition, exert pressure on PMSCs to be more militarily effective, we can better highlight instances when accountability is at its most vulnerable point. Once we identify such moments, we can begin to propose new ways to curb opportunistic behavior among PMSCs that operate on the market today and attach at least some value to their reputation as they hope to survive in a competitive environment. Our goal is to build on existing proposals to make accountability a more tangible reality for private military and security providers. This book shows how to harness the power of the market to shape the behavior of one type of PMSCs, international corporations, to increase their incentive to invest in material/individual capabilities and enhance their corporate professionalism—factors that contribute to military effectiveness.

Initial data we presented in this chapter demonstrate the growing presence of PMSCs in major and minor civil wars since 1990. While PMSCs have delivered a myriad of services, they have most frequently either engaged in combat or provided combat assistance to the governments' troops, thereby making them important players in shaping conflict dynamics. In the next chapter, we delve into the idea of local competition faced by PMSCs intervening into civil wars. Why does this competitive force matter? And how does local competition affect PMSCs' military effectiveness in contributing to the shortening of wars?

# 3

# Local Competition and PMSCs' Behavior in Civil Wars

As the civil war raged in Sierra Leone in 1995, the government turned to Ghurka Security Guards (GSG), a PMSC, to train the country's Special Forces and combat the rebel movement, the Revolutionary United Front (RUF). While GSG, a sole security provider to the government of Sierra Leone, was initially successful in securing gains for the government, the company refused to engage the RUF when several of their men were killed in an ambush, thereby forgoing some of their contractual responsibilities. Yet a few years later, GSG had become known for taking greater risks and engaging the enemy in Afghanistan, where the company operated as one of several PMSCs. Why is it that GSG had undertaken risks in Afghanistan yet ultimately failed to be militarily effective in Sierra Leone, thereby pushing the government to replace it with Executive Outcomes, then a rising giant in the private military and security industry? Existing literature assumes that PMSCs either improve conditions in wars or their impact is detrimental and thus governments need to devise better tools to regulate the industry. While governments such as that of the US have improved their ability to monitory the industry through contract clarity and ANSI/ASIS or ISO certification, on the field accountability has remained most challenging, even more so for weak states. Yet some companies have nevertheless become more accountable to the clients on the job by adapting strategies that increase their military effectiveness. GSG was taking more risk to complete its mission in Afghanistan than it did in Sierra Leone in 1995. What accounts for the change in their tactic and how does the change in behavior shape civil war dynamics? More generally, how do market dynamics shape PMSCs' behavior in civil wars?

The argument we advance is that PMSCs face two major competitive environments: the rise in general competition at the industry level or the unprecedented expansion of the industry marked by low entry barriers, and local competition they encounter in a given conflict. These market forces

*Private Militaries and the Security Industry in Civil Wars.* Seden Akcinaroglu and Elizabeth Radziszewski, Oxford University Press (2020). © Oxford University Press. DOI: 10.1093/oso/9780197520802.001.0001.

have greatly affected PMSCs' strategies and hence their behavior in conflict zones. Much has been said about the growth of the industry, yet few have explored how PMSCs have adapted their strategies to the evolving market challenges and how these changes have impacted their behavior in civil wars. Our focus in this chapter is on how local competition among military and security providers in a given conflict zone affects PMSCs' ability to be militarily effective in fulfilling the government's objectives in ways that contribute to conflict termination.

To examine the impact of PMSCs on the termination of civil wars by focusing on market pressure that private military and security providers face, we begin with several assumptions. First, PMSCs are rational, self-interested actors whose main goal is to secure profit. Put simply, PMSCs operate as corporate ventures and their continuous survival depends on their ability to maintain financial gain.[1] This makes PMSCs vastly different from public forces, in that the latter serve a single state while the former can boast a diverse clientele of states. Second, considering that PMSCs provide military and security services, they benefit from an environment where threats are present. Whether they help weak governments in Africa bolster their public armies' efforts to thwart rebel incursions and terrorist attacks, or guard sites in Iraq, the existence of security threats creates the market for PMSCs' services. This is not to say that PMSCs only operate in active conflict zones; in fact, they engage in domestic peacetime training and development that furthers countries' regional and global security agenda even in the absence of imminent threats.[2] Yet the work of many international PMSCs has focused on logistical support to troops, military and police training, intelligence gathering, combat assistance, and occasionally direct engagement. The conflict environment then ensures a steady supply of income to most PMSCs. Consequently, most PMSCs benefit from the presence of insecure and conflict-prone environments. As such, there is a potential gap between PMSCs' interests and those of the clients they represent. Cockayne (2007),[3] for example, argues that the PMSC-client relationship can be best examined in terms of a principal-agent framework, where the PMSCs act as the agents of the principal or the warring party but pursue their own interests instead of those of the principal. The principal, after all, is in most cases interested in terminating the hostilities and defeating the opponent as swiftly as possible.

In limited instances, leaders might be motivated to initiate or intensify conflict to demonstrate competence and score political benefits, an idea captured in the diversionary theory of war.[4] It has been argued that in order to discredit domestic opposition, Assad released thousands of jihadists from Syria's prisons to spread chaos and present himself as the defender against ISIS extremists.[5] However, evidence that examines diversionary tactics at the international level shows that in most cases extending the conflict backfires.[6] The rally effect is usually a short-term phenomenon and prolonged diversionary attempts create perceptions of leaders' incompetence. Finding a target for diversionary tactic, whether internationally or domestically, is not always easy. Consequently, it is likely that in the majority of cases leaders will benefit most from temporary rather than extended violence and thus seek a shorter duration of war. While the government hopes for effectiveness and quick termination of conflicts, it may, however, get something different: a company that prolongs its contract and does just enough to convince the government of impending victory if only it could stay longer to finish the job. The essence of the principal-agent problem is that governments are often unable to effectively monitor PMSCs' behavior and this creates an opportunity for corporate actors to exacerbate threats or engage in fraudulent practices that may extort resources from the government. When contractual obligations are vague, PMSCs may push for tactics that may appear as successful initially but ultimately contribute little to overall military effectiveness needed to terminate the war. For example, excessively aggressive interrogation tactics may be a way to gain quick access to intelligence, but they would alienate the locals and impede effective intelligence gathering in the long run.

Third, we assume that even though PMSCs benefit from the conflict environment, those with a corporate structure also attach value to their long-term reputation. If the client discovers that a company deliberately sent a limited group of experts than promised or failed to respond to changing dynamics in a conflict, a loss of contract and poor image are likely to follow, thereby threatening the business's existence. In some cases, prolonged and bad publicity could undermine the companies' abilities to attract new clients. To a lesser significance, yet still relevant, is also the potential to lose valuable connections with weapons and equipment suppliers who might sever relations with companies marred by scandals and poor reputation.[7] The significance of long-term reputation is especially relevant in the post–Cold War environment that has seen an increasingly crowded marketplace of security professionals delivering ever more sophisticated services.[8] For PMSCs that have limited financial options, such as those in the UK, for example, reputation is vital for securing public contracts.[9] This general trend at the industry level has contributed to greater competition among security providers.

In the past, limited emphasis on norm emergence in the private military and security industry and the relative newness of industry meant that fewer companies were concerned with the reputational consequences of their behavior. In this environment, it was easier for companies to neglect humanitarian concerns or rely on poor-quality equipment as long as doing so provided short-term gain in the battle. As the industry began to expand, however, and negligent behavior of some providers has been documented and condemned by international audiences, companies have become more sensitive to accountability and reputational costs. In the immediate post–Cold War environment and even more so today, reputational costs for PMSCs are linked to unethical behavior and weak military effectiveness increasingly attributed to behaviors such as indiscriminate use of force that alienates the locals and encourages them to side with the insurgents. This is usually true for companies established along corporate lines, where emphasis is placed on developing strategic mission for long-term growth. Among those, the acknowledgment that reputational concerns matter for the company's success eventually kicks in, sometimes through observation of changing trends in the industry, while at other times the need emerges after poor behavior is publicly revealed. Such companies may still survive but doing so takes serious rebranding.[10]

Reputational concerns are more limited in the context of the so-called fly-by-night companies that are temporary in existence.[11] Such companies are not designed with long-term corporate development in mind and thus may exhibit little or no interest in upholding practices that improve their reputation. For example, they might send employees into a conflict zone with poor technical skills or rely on low-quality equipment without considering how such practices would affect their ability to renew a contract. Overall, we assume that most companies organized along corporate lines exhibit some concern for long-term reputational costs in light of growing market pressure while recognizing that temporary companies would be less susceptible to such forces. Based on the three assumptions about the interests of private military and security providers, we examine how the environment in which PMSCs operate, the market pressure they face, affects the companies' effectiveness in shifting the balance of power in favor of the government and thus in helping to terminate the war.

## Local Competition, Effectiveness, and PMSCs' Accountability in Civil Wars

The idea of local competition, or pressure that companies face as the number of providers that the government hires in a given conflict increases, is rarely considered in the PMSC literature.[12] Yet the importance of local competition is likely to matter significantly given that governments have been increasingly relying on the services of multiple providers, as seen in the cases of Iraq, Afghanistan, and Zaire (now Democratic Republic of the Congo). A particularly notable example of this competition was evident during the Iraq war. In 2007, there were more private contractors from multiple companies working on the US payroll there than US troops.[13] From 2004 to 2019, 137 international PMSCs provided a variety of services in Iraq to the US government from logistical support to convoy protection.[14] Figures 3.1 and 3.2 show the presence of PMSCs across conflict years for both major and minor civil wars from 1990 to 2008. Figure 3.1. demonstrates that in 1992 there were three conflict years with PMSCs operating in major civil wars and those interventions were all cases where only one PMSC was hired by the governments. In 2006, there were three conflict years where PMSCs were present; two cases involved governments working with just one provider and one case in which the government relied on multiple providers. Both figures illustrate that

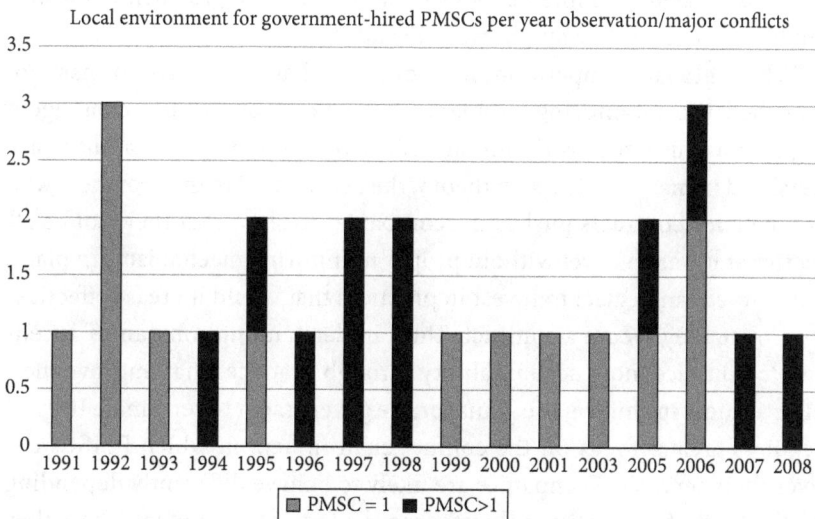

Figure 3.1  Local competition for government-hired PMSCs in major civil wars

Local environment for government-hired PMSCs per year observation/minor conflicts

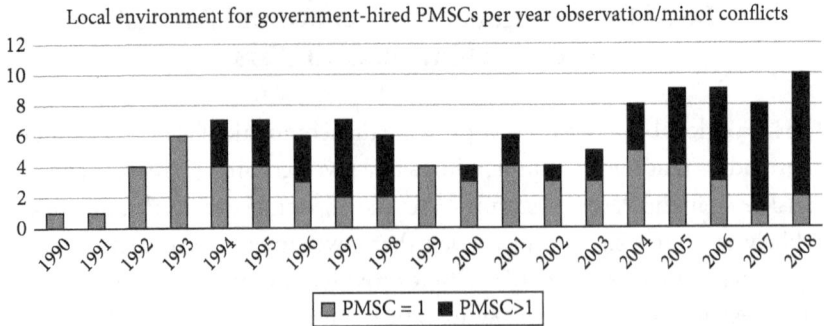

**Figure 3.2** Local competition for government-hired PMSCs in minor civil wars

the presence of multiple providers in conflict years increased starting with 1994. Before 1994, PMSCs operated alone in the market and faced an uncompetitive local environment. This dynamic has shifted and reflects the industry's expansion. After 1994, governments began to have more options and could hire multiple providers because more companies had appeared on the market. In particular, the trend to hire multiple providers has dominated in minor conflicts since 2004 (Figure 3.2). The overall pattern from 1990 to 2008 shows the variation in the level of local competition, with more competitive environments dominating in post-1994 years.[15] Governments have relied on PMSC services in 15 major conflicts, with eight of these conflicts being associated with interventions by more than one provider. For minor conflicts, the respective numbers are 35 and 19.

While global competition for contracts has increased, it has not banished the monitoring problem at the heart of the principal-agent logic, and thus issues with military effectiveness and accountability have persisted in many contexts. In theory, the existence of many providers who compete for contracts pushes the companies to show that they can be effective at lower cost, yet without proper monitoring mechanisms in place, companies can neglect to invest in practices that would increase effectiveness once they secure a contract. Thus, understanding companies' incentive to embrace more accountability through practices that improve their effectiveness in shifting the gains for the government to terminate the war requires greater focus on the conflict environment in which PMSCs deliver their services. Companies are likely to behave differently depending on the level of competition they face in the local environment where they work for the client.

Consider the case when the company encounters limited competition in a given conflict environment. When a company is the sole provider of services, it has more opportunities to use lower-quality equipment or less-experienced employees than promised because the oversight mechanism is usually low in such cases. It is also less likely to be responsive to changing conflict dynamics, as was the case with Ghurka Security Guards in Sierra Leone when the company refused to engage the enemy when some of its men were killed. In Chapter 2 we argued that military and individual capabilities can affect military effectiveness, and companies that deliberately utilize poor-quality equipment or employ technically unskilled workers might score lower on this dimension of effectiveness. We also argued that corporate professionalism as evident in the culture of integrity that reduces instances of fraud and human rights abuses is a contributing factor of military effectiveness associated with quicker termination of conflicts. Companies that operate as sole providers of military and security services in a conflict zone have a lower incentive to be fully transparent and follow through with practices that reflect corporate professionalism, even if they demonstrated their capacity to do so in the bidding stage.

First, creating a credible culture of transparency may be damaging to a company that deliberately seeks to extract additional resources from the government while leaving no trace of illegal activities. Second, building a culture with zero tolerance for corruption at the lower levels of the hierarchy among the unit commanders and other employees in the field is costly and therefore inconsistent with the companies' short-term goal of making profit with limited investment. As sole providers of a service in a conflict zone, companies are more likely to get away with fraudulent practices without being caught; therefore, not only is there a high temptation to cheat on that dimension of military effectiveness but there is also limited interest in spending money to hire men who have been vetted for the quality of their character and to foster the growth of norms against corruption for all employees. As we argued in Chapter 2, fraud limits military effectiveness in a conflict zone because the government may lose valuable resources that are extorted by a PMSC and which could be used to finance additional missions against the insurgents. When limited military effectiveness occurs and progress against insurgents is slow, companies then exhibit low levels of accountability to their client because they are not able to shift the balance of power in favor of the government decisively enough to end the war. Yet they can find a way to preserve their reputation by claiming that success in complex conflict environment requires more time and contract renewal. Being a sole provider in a conflict zone in an on-the-field environment of limited oversight encourages such practices, with negative consequences for conflict termination.

A former Pacific Architects and Engineers employee noted the limitation of having only one company holding various contracts in one program and the benefit of competitive pressure. "The good thing of having one company is that everything is managed under one roof. You would think that is a good thing, but if for some reason the company does not place good US personnel as in-country managers, then you have a serious problem that significantly affects the effectiveness of the program. If you have more than one company holding various contracts in one program, then you have the cut-throat business when it comes time for re-bids. If the companies have an on-site competitor, which they usually do for large programs, then certainly it is something that keeps them in check to do the job as best as possible and to recruit/hire great personnel."[16]

There are also lower incentives to winning the hearts and minds of the locals in the absence of local competition. Commitment to humanitarian laws, as evident in population-centric COIN, is often connected to the government's military effectiveness, the idea we explored in Chapter 2. The locals often consider PMSCs that work for the government as an extension of the government's armed forces and thus their respect for human rights is likely to affect the government's ability to defeat the enemy. In her research on PMSCs' military effectiveness in Iraq, Dunigan (2011) concludes after interviewing numerous US military personnel that "PSC [Private Security Contractors] noncompliance with legal and ethical norms of just war has a detrimental impact on military effectiveness."[17] Even when the military pursues a more accommodative tactic toward the locals, PMSCs' violations of humanitarian laws can undermine the military's mission when the latter has to clean up the wrongdoings committed by contractors. Carmola (2012)[18] also notes that risky behaviors, such as the fatal shooting of two Afghan security guards by PMSC employees in 2009, exacerbate anti-American feelings and hinder the war efforts. In contexts where PMSCs are hired to do the bulk of the work, they are also likely to be considered as an extension of the government and their abuses will likely be linked back to the state. The consequences could be dire for the government's ability to secure substantial gains against the enemy and for conflict termination. As locals begin to side with insurgents and withhold vital information on enemy movement, insurgent resolve is likely to increase.

Reports about armed contractors using excessive force, intimidation, and threats while deployed in conflict zones have come to light over the years. Such acts are more likely to occur when the monitoring mechanism is low, as is the case in conflicts where contractors operate as sole security providers. Although we expect that international PMSCs might exhibit greater sensitivity to international norms than mercenaries and ad hoc companies, they are not immune to committing such abuses. Civil wars are more likely to last longer when PMSCs operate as sole providers because the absence of local competition, the limited market pressure at the micro level, reduces the incentives to adhere to international humanitarian standards once the contract is secured. Depending on the government and conflict zone, there might be some to no oversight of PMSCs' interactions with the locals. Yet why would PMSCs choose to violate humanitarian laws in the first place, especially if they are military experts with the potential to understand the benefits of population-centric COIN in helping governments win the wars? Wouldn't the benefit of population-centric COIN push them to uphold such standards?

Addressing these questions requires a return to the initial assumption about PMSCs' goals, which are predominantly about securing profit. PMSCs may sometimes deliver intermediate victories for the government by using indiscriminate force, a strategy that is cheaper to implement than fighting insurgents by getting the locals on your side. The latter requires more skillful employees and greater levels of responsiveness and creative thinking to accomplish. If a government hires a company for a very specific mission, such as keeping resource-rich areas away from insurgents' control, it may also be satisfied when it sees that insurgents have abandoned the region. The company has secured a victory and leaves. Yet while aggressive tactics may secure the battle, they rarely win the war for the government.[19] The company fulfilled what it promised, yet it was not accountable to the government in the broader context of winning the war. When governments hire companies for a specific mission, there is an expectation that these missions will contribute to the government's victory and conflict termination. While some governments may be indifferent to the tactics, most care about securing a victory that lasts. In the absence of monitoring mechanisms, as is the case in a non-competitive local environment, PMSCs are less likely to adhere to international humanitarian laws because they can claim that they have delivered a victory to the government and thus have been effective in fulfilling the mission they were expected to do. In reality, however, they are not highly effective and thus fully accountable to the client when the tactics they use to fulfill a specific mission undermine the government's overall war strategy. In this case, they have merely sold an illusion of military effectiveness to the government. It may be only months after the company departs that the victimized population joins the rebels, the fighting expands, and the disastrous consequences of PMSCs' tactics become more apparent.

Carmola (2012)[20] argues that PMSCs have been in the business of taking on risks that military needs to avoid. She notes that PMSCs do not purposely engage in problematic behavior but might do so because there is a certain expectation that they, in fact, should. When such PMSCs operate alongside the military with lower tolerance for risk taking, there could be confusion about strategies in already complex conflicts, which could negatively affect the balance between tolerance for risk and the establishment of trust from the locals. This argument contrasts with our earlier point about a more conscious and deliberate decision to forgo humanitarian interests due to costs associated with investing in the necessary training and creative thinking required to build local trust. While there are instances when PMSCs might be expected to do the "dirty" job, in many cases they have influence over strategy development or work for clients who either refrain from routinely communicating such an expectation or who openly oppose it.

Deterring opportunistic behavior—opportunism that we argue is a key factor behind risky, indiscriminate use of force that harms civilians—is critical for greater accountability of PMSCs and because being a sole provider of services in a conflict zone incentivizes limited accountability, the issue of monitoring deserves a more careful look. Weak monitoring mechanisms in conflict zones are a problem for most states, but are especially acute in weak states. Weak states have limited institutional capacity and are marred by inadequate rule of law, inefficiency, and vulnerability to crises.[21] As such, the governments of weak states may be unaware of the extent to which a company engages in fraudulent behavior, for example, by manufacturing threats to stay afloat with potentially devastating consequences on the government's ability to terminate the war quickly. Many governments in conflict zones that turn to PMSCs in the first place face a challenging enemy and may have only a limited idea about the necessary approach to defeat the opponents they were unable to eliminate on their own. Avant (2005),[22] for example, argues that because of their expertise PMSCs have affected states' monopoly over the use of force. At times, governments may allow a PMSC to shape their expectations for the type of acceptable results because of their limited ability to monitor the situation on the ground and sufficient expertise to evaluate the companies' strategies. This allows a PMSC to be less militarily effective than it could be without facing reputational consequences. As long as the PMSC is able to deliver some results to the government, it can justify why it needs to stay in the conflict zone for six years, for example, as opposed to only three.

Many conflicts in weak states such as those in Africa are difficult to resolve—fighting in the DRC or Somalia are some examples of cases where conflict termination has been elusive for years. There is often an expectation that working in such conflict zones is especially challenging. In the Congo, infrastructure is in such dire state, with parts of the country impenetrable, that during the 2011 presidential elections voting ballots were delivered to remote regions by helicopters.[23] Because operating in some conflict areas is complicated, a company that performs less effectively yet helps the government make limited progress against the enemy is less likely to face reputational costs. Given the governments' weak monitoring capabilities, the expectation of dealing with a strong enemy, and the potential for a PMSC to define new security threats to governments, a single PMSC could appear as militarily effective even though its ability to help the government secure a quick victory might be questionable. In other words, there exists a discrepancy between actual and perceived level of effectiveness.

It is important to note that being a single security provider in a conflict zone does not eliminate all pressure on companies to perform. A company still has to deliver something so its contract is not terminated, but it need not deliver a lot to stay afloat. As long as a PMSC can justify limited gains or frame those gains as effective in shifting the tide of war, a company can maintain its reputation. As we have argued earlier, governments in such cases might not be able to determine whether the PMSCs' effectiveness could have been better than what they delivered due to weak monitoring capacity and limited understanding of the unfolding dynamics in the field.

This opportunistic behavior is highly dependent on the level of competitive environment in which PMSCs operate and the variation in oversight that such an environment generates. Unlike cases where PMSCs operate as single security providers, an environment where local competition is low, an increase in the number of PMSCs creates more pressure for companies to improve their military effectiveness and thus become more accountable to the client. Porter (2008)[24] argues that the entrance of new competitors in the industry is one of the critical environmental threats that affect organizations' attractiveness and strategy. In the case of PMSCs, the threat stems not only from the establishment of new firms with their diverse and sophisticated menu of services, but also from the presence of one or more players in a given conflict environment, that is, the entrance of additional actors into the local market. Competition allows the government to compare companies' performance and this transparency pushes each one of them to set the bar high in terms of reporting their military effectiveness. This process is undermined in contexts with a high level of specialization in service delivery, yet this is rarely the case for PMSCs. Most PMSCs have a broad portfolio of services to adjust to different needs of clients and environments and thus face a high level of competition from each other.[25]

How can governments, however, determine whether the success of one company is real and not merely manufactured to offset the competitor? Put simply, the complexity of conflict dynamics does not diminish with the presence of multiple players; on the contrary, in some instances the governments rely on multiple companies because the conflict is complex and the monitoring in such situations may be especially challenging. Consequently, while having more providers allows the government to compare their effectiveness, it does not always help the government determine whether reported gains are credible and whether a given company could have been more effective in shifting the tide of war had it adapted better strategies (e.g., gaining the trust of the locals) or refrained from opportunistic behavior associated with fraud. Instead, governments must rely on companies to monitor each other to determine the true nature of their gains against the enemy.

In his work on principal-agent problem, Varian (1990)[26] shows that mutual monitoring, where agents alter other agents' costs and benefits of engaging in undesirable acts, can provide a particularly cost-effective way for the principal to scrutinize the behavior of its agents. Specifically, rival companies should have incentives to monitor each other's misconduct and lack of competence. The presence of competitors introduces the monitoring element that weak states with limited resources are traditionally missing and strong states struggle with, albeit to a lesser degree. Human rights organizations and other civil society movements, at times, receive anonymous tips from industry insiders or employees, the so-called whistleblowers, about witnessing abuses against the locals or having knowledge of such abuses being committed.[27] When local competition increases, the stakes are simply higher for companies that are not effective in helping the government secure gains against the insurgents. The fact that several companies operate in a given conflict means that even a government with seemingly weak monitoring capacity can enhance its control over the agents and strengthen its bargaining power.

The presence of several companies thus creates a potential way to monitor PMSCs in conflicts and may help reduce behaviors that undermine military effectiveness. Consider, for example, how competition affects a company's use of military and individual capabilities. Knowing that they could be monitored by competing firms, PMSCs might choose to hire better-skilled workers in a given conflict or rely on high-quality equipment. Indeed, when in 2004 Spicer's British company, Aegis, came up with a solution for the DoD to let a private security company police other contractors and got the $293 million contract in Iraq, its rival in the local Iraqi market, DynCorp, was the first one to file a protest with the Government Accountability Office asking for Aegis's credentials and competency to be rechecked given its murky past. Knowing the incentive of its competitor to discredit its services, Aegis had to nail the job. Despite cutting costs while bidding for the contract, as its co-founder Dominic Armstrong said, the company installed tracking devices on contractor vehicles, monitored and coordinated its competitors' movements from Baghdad's Reconstruction and Coordination Center (ROC), and reduced the clashes between PMSCs. In a January 2009 report, the Special Inspector General for Iraq Reconstruction (SIGIR) noted that "the firm had tracked 55,000 contractor movements and reported nearly 400 infractions of the rules, including eighty by its own staff—a performance the agency rated as satisfactory to outstanding."[28]

Potential monitoring by rivals could also put pressure on PMSCs to send employees into the field who demonstrate a high level of responsiveness and innovative thinking that may be required to deal with rebels who shift tactics or alliances. In doing so, the company could be more effective in securing gains for the government quicker. It is entirely possible that in the absence of competition, a company could shirk on these dimensions yet still achieve some gains for the government. However, as we have argued throughout this chapter, military effectiveness is about achieving gains quickly and in ways that do not jeopardize the government's ultimate goal of winning the war. Competing firms can also push each other to adapt higher standards of integrity in a conflict zone when it comes to minimizing fraudulent practices such as low staffing or taking bribes from militias. Being caught committing fraud by a rival company creates risks of exposure, blackmail, negative publicity, and severe reputational costs.

The same is true in the area of human rights abuses. In their study of Nordic businesses, Christiansen, Hansen, and Poukka (2015)[29] note, for example, that when asked about key drivers influencing companies' commitment to human rights, peer pressure came at the top while regulatory pressure was rarely cited. This competitive factor played a role in EO's embrace of a more population-friendly approach to securing gains for the government. Recognizing this phenomenon, Nic Van Den Bergh of Executive Outcomes (EO) noted, "The fastest thing that would get us out of business is human-rights violations."[30] EO understood the importance that winning the hearts and minds could have on delivering a victory to the government and enhancing its reputation for military effectiveness. Thus the company placed restrictions on the use of force as part of a strategy to build good relations with the local population in order to receive intelligence on the rebels' movement.[31] EO's pilots flew their helicopters close to the ground rather than choosing a somewhat safer but less effective tactic of flying higher and escaping enemy fire. Being close to the ground enabled the pilots to better distinguish between the rebels and the civilians and engage in more targeted shooting.[32]

Even if the client is less concerned about human rights abuses from an ethical perspective or hires PMSCs to gain plausible deniability and avoid oversight, a rival company may still have an incentive to exploit a competitor's weak adherence to international humanitarian laws. A rival company could shed light on the risks associated with indiscriminate violence as a strategy of winning the war and in doing so discredit the competition. A shrewd competitor might benefit from exposing a rival company's violations in future contract bidding, as companies have done in the case of Iraq. Therefore, the incentive to report on humanitarian abuses in a given conflict is not solely shaped by the extent to which the current client prioritizes it. Put simply, even if the client is openly indifferent to humanitarian laws and/or seeks to gain plausible deniability that lowers its expectations of PMSCs' adherence to such laws, self-regulation in the area of human rights can still prevail as long as companies face a competitive local environment.

The possibility of being monitored by other service providers creates what economists often refer to as informal regulation,[33] a phenomenon that could complement existing efforts to regulate the industry. How does this form of regulation, however, differ from formal regulatory mechanisms currently propagated as solutions to improving PMSCs' accountability? Formal regulatory mechanisms have made a significant contribution in bringing adherence to international humanitarian laws to the forefront of the debate about the dangers of using private contractors in conflicts. Companies can increasingly focus on monitoring a competitor's treatment of the locals—for example, by reporting to the client, NGOs, or the media when initial abuses occur and prompt a change in the strategy from the alleged violator that results in greater awareness of population-centric COIN—because initiatives such as the Montreux Document have set expectations for the industry and clients in the area of good practices. Because clients are now asked to explicitly demand that PMSCs uphold the standards of international humanitarian laws and to work only with those companies that show commitment to such laws, companies have a standard by which to hold each other accountable. Monitoring each other on fraudulent practices or poor military and individual capabilities has been more obvious, yet using adherence to international humanitarian laws as a monitoring standard has gained traction even more so over the years in part due to these international initiatives.

Yet the remaining challenge is to increase compliance with these norms for an industry lacking in the area of transparency, especially when it comes to PMSCs operating in complex conflict zones. Here is where informal regulation enters the picture. While formal regulatory mechanisms set the important ideals for accountability, they do not guarantee that PMSCs will be motivated to adhere to these principles on the job. In the end, PMSCs are profit-oriented businesses. Their priority is to secure financial gains, so increasing accountability must be intertwined with this goal. The pressure to get ahead of the competition creates an incentive for companies to monitor rivals and potentially exploit their human rights abuses for competitive gain. It is also a factor that can lead companies to make different choices from the start about the type of employees, equipment, and strategies they will use in a given conflict zone. Harnessing the power of the market by creating a competitive environment in a given conflict is thus a way of pushing companies to act upon the humanitarian standards outlined in the formal codes of conduct beyond the bidding phase. The significance of informal regulatory mechanisms, such as the presence of local competition we describe here,[34] is that they work because they do not codify or demand certain behaviors. Companies simply adjust on their own to the changing environment of competitive pressure.[35] Drutschmann (2007)[36] points out that: "unlike formal regulation, such as law and contractual specifications, which relies on 'codification, (intrusive) monitoring and safeguards' in the form of explicit enforcement systems, informal regulation is generally uncodified, relies on cooperative monitoring and has no explicit safeguards. It emphasizes the importance of elements, such as trust and social pressures, which emerge from the social interaction that underlies each economic interaction in order to achieve regulatory control."

Companies that expect formal regulation from third parties joining them in the field to monitor their behavior are likely to improve their effectiveness, especially in the area of aggressive treatment of the locals, but this mechanism has some limitations. In the case of Iraq, for example, there were shortages in qualified contracting US agents who could systematically monitor PMSCs. In weak states, contracting agents may not even accompany PMSCs on their mission. By contrast, the idea that a rival company may or may not be tempted to report on systematic abuses creates a perception of pressure about being caught that could contribute to existing monitoring efforts or compensate for the lack of such efforts in some contexts.

Research shows that deterrence of criminal behavior can succeed even when actual monitoring and punishment is not objectively high.[37] As long as there is a general perception that monitoring is a possibility and negative consequences for pursuing opportunistic behavior exist, there is a strong likelihood that individuals will monitor themselves. Although there is some disagreement as to whether effectiveness of deterrence depends on objective or subjective certainty, findings point to the importance of perceptions of risk over actual risk of being caught and punished.[38] For perception of risk to exist, there must be some credibility behind the idea of getting caught and the ensuing punishment. The credibility of punishment for failing to honor international standards has increased for international PMSCs in part due to the myriad of international and industry-wide efforts—such as those laid out by ICoCA and the Montreux Document—to raise the agenda of ethical behavior and to create an expectation among clients and the industry about meeting these norms or risk reputational costs that could result in loss of contracts to rival companies. Such was the case, for example, with ArmorGroup North America in 2009 when revelations of misconduct while on contract for the US government guarding the embassy in Kabul led to the cancellation of the company's contract and its replacement with a rival company, Aegis.[39] Employees in the field can also observe the existence of negative consequences for those employees who engage in opportunistic behavior. The Nisour Square shooting incident, for example, culminated in 2014 in guilty verdicts on the charges of first-degree murder and manslaughter for four former employees of Blackwater.[40] In many instances, companies have adopted a zero-tolerance policy pertaining to human rights violations and fraud. During several operations by PMSCs in West Africa, employees have been removed from the field immediately and fired.[41]

But can PMSCs expect to be caught if engaging in an opportunistic behavior? For that to occur, there must be an expectation either that NGOs will report on such behavior or that the employees of one company could expose an employee from another company for systematic abuses that might be linked to lower military effectiveness. Such an expectation exists, although the motives for monitoring are more diverse at the employee level than at the CEO level. While competition is cited, there are also other factors motivating peer reporting in the field. Unlike CEOs who might adopt specific strategies, such as improvements in the vetting of employees, to ensure they can compete on military effectiveness with rival companies in the same conflict zone, competition is a lesser concern among the employees in the field. Instead, concerns about safety posed by unethical behavior of men from different companies are an issue.[42] Opportunistic behavior may, at times, put other men in danger, thereby motivating a response, something that becomes a more significant concern when multiple companies operate in the same conflict zone. In a competitive local environment, the likelihood that risky behaviors from employees of one company could have a negative spillover on employees of others increases simply because there are more players involved with the possibility of engaging in such behaviors. As captured in the words of one PMSC employee who worked in Liberia and Sierra Leone, "[angering] the locals has an effect on all the whites," and so concerns about spillover effects are taken seriously in the field.[43] Consequently, when multiple companies are present and employees recognize the possibility that their actions can have an impact on many additional actors, they are more likely to expect that players beyond their own company would report risky behaviors, thereby generating pressure to self-monitor.

In some cases, employees of other companies are also motivated to report on others to uphold ethical standards. Whether the latter reason is the true motivator is not as vital as the perception that it could be. As more companies are putting greater emphasis on fostering a culture of professionalism and disseminating information about their undertakings, as was the case with DynCorp, which distributed a pamphlet about its commitment to ethical standards not only to NGOs but also to their competitors,[44] employees of the industry in general are likely to anticipate more determined effort by at least some employees to uphold these standards in the field through monitoring. Employees in the field are aware of the possibility that rivals, and their own co-workers, may tip the media or reach out to NGOs. Such was the case, for example, when ArmorGroup North America's employees contacted Project on Government Oversight (PROGO), an independent nonpartisan watchdog, in 2009 with revelations of flawed security at the Kabul embassy. While some of these opportunities for reporting abuses are likely to exist regardless of whether one or more security providers deliver services to the client, there is likely a perception that motivation for monitoring is greater with the addition of multiple companies. It also means that there are now additional players in the field that are knowledgeable about military and security tactics and that could also work jointly with other PMSCs as part of a grander mission. These new players are thus in a unique position to access hard-to-get information. One of the limitations of increasing the presence of the media and NGOs to improve monitoring is that such actors, while helpful in this regard, may not necessarily gain additional access to sensitive information about the effectiveness of tactics and behavior that is visible to people with more sophisticated military training and more continuous presence in the battle zone. The media and NGOs play a significant role in disseminating such difficult to obtain information, yet they must first gain access to it for the mechanism of accountability to work. Here is where additional presence of PMSCs as security experts provides value.

Consequently, self-monitoring is likely to be greater in more competitive settings where the number of security providers increases and with it the possibility that more qualified people might be willing to report practices that might be detrimental to PMSCs' military effectiveness. Lastly, self-monitoring as a response to a competitive environment begins with a corporate decision to hire more qualified, professional contractors, who are thoroughly vetted and thus less likely to engage in opportunistic behaviors to begin with. In the absence of a competitive local environment, companies may simply have limited incentives to pre-select those individuals who are risk-averse when it comes to opportunism.

Nevertheless it is worth exploring an alternative possibility to the mechanism of informal monitoring discussed here, one that envisions collusion among PMSCs, a secret pact to turn the other way when witnessing mismanagement, a phenomenon that could spell a demise of informal regulation's deterrent mechanism. According to one industry insider, collusion at the company level is unlikely to occur, mostly because companies are more likely to treat each other as competitors than allies.[45] This point is further emphasized by McFate (2014)[46] in his case study of DynCorp's and PAE's reconstruction work in Liberia. Rather than collude against the client, in this case the US government, the companies worked separately, with PAE later being assigned the role of "peer reviewer" of DynCorp's work. Secret deal cutting is risky for companies that seek to safeguard their reputations. When failure by one company to reveal an abuse of another occurs, it does so when an employee refrains from reporting misconduct due to personal connections to an employee from another company, as might be the case if, for example, former comrades working for the British Special Forces are employed by competing companies. Personal networks can in this regard serve as a potential hindrance to whistleblowing.[47]

To what extent do personal networks undermine the accountability model? Literature on whistleblowing argues that people are willing to forgo protecting their personal connections when they perceive the offense to be serious and in violation of ethical standards.[48] We expect that PMSC employees might be more motivated to disregard personal relationships when offenses are in fact major, such as selling secrets to the enemy, abusing civilians, or arriving in the field unprepared to conduct the mission in an attempt to cut expenses and potentially jeopardize the safety of others. In instances where PMSC employees engage in minor offenses, for example, driving recklessly or showing disrespect toward local culture, the literature indicates a higher incentive to turn the other way, limiting the efficacy of self-monitoring. It is difficult at this point to assess how detrimental these low-level offenses are on the overall success of the military mission. While Cotton et al. (2010)[49] argue that even such offenses may alienate the locals, it is not clear whether the effect is temporary and localized. By contrast, higher-level offenses are much more likely not only to be publicized but also to be used by the enemy as propaganda for larger audiences, in all likelihood generating a much greater negative effect on the mission's overall success.

The opposite problem to consider is that companies not only refrain from colluding practices but may try the opposite, present false accusations against rival companies, leaving the client confused and faced with even greater information asymmetries than before the companies entered the picture. Defamation, which we discuss further in the next chapter, however, also carries enormous risks. For international Western companies there is an opportunity to resort to legal means in their own countries against slanderous accusations concerning their work in another country. This means that companies tempted to present false information against a competitor must assume the possibility that a legal process could be involved, one that could very well tarnish their own reputation for spreading rumors and false information. As such, reputational concerns could mitigate practices that increase informational discrepancies and would be detrimental to the clients' interests.

With the presence of informal monitoring, PMSCs are likely to behave in ways that increase their effectiveness and accountability toward the client. In turn, if each government-hired company displays high level of military effectiveness in the area of skill and corporate professionalism, the government is likely to gain an advantage over the rebels. This either should enable the government to achieve victory or would prompt the rebels to negotiate. Overall, when local competition is high, the companies' intervention is likely to contribute to a swifter termination of the conflict.

While our focus is on PMSCs' interventions on behalf of the government, it is worth noting that on rare occasions, rebel organizations have sought assistance from PMSCs. In those cases where PMSCs intervene on behalf of the rebels, we expect their involvement to shift the balance of power in favor of the rebels. Rebel organizations with funds to hire a PMSC are likely to have significant resources to wage an armed conflict, thus presenting a reasonable challenge to the government. Incoming assistance from PMSCs could tip the military balance in favor of rebel victory because groups that can afford to hire such companies are unlikely to be weak in the first place. There are several ways in which PMSCs can help the rebels achieve victory. Besides aiding in combat, PMSCs can also become indirectly involved when they serve as liaisons in weapons trade. In fact, anecdotal evidence suggesting that PMSCs facilitate connections between sellers and buyers has prompted the international community to devise measures that would monitor the firms' activities.[50] By providing better military capabilities to the rebels, PMSCs could be effective in helping the rebels make substantial gains against the government. Greater military effectiveness of PMSCs hired by the rebels is expected, particularly in situations when the rebels rely on multiple companies for services. Just as we argued that competition could push PMSCs to greater effectiveness and accountability when supporting the government, so should the same be true in the cases of intervention on behalf of the rebels. More competition reduces the principal-agent problem in a similar manner when rebels are the clients.

Our data reveal, however, that international PMSCs are reluctant to support the rebels and local competition is rare for PMSCs working for such actors. The rebels relied on PMSCs' assistance in 7 of the minor conflicts, and in 6 of the major conflicts. In only 3 of these major and minor conflicts do we observe multiple providers working for the rebels. Figures 3.3 and 3.4 show the number of PMSCs across conflict years for minor and major conflicts. In both types of conflicts, 1998 is the last year in which we see multiple providers hired by the rebels. One of these conflicts had erupted in Sierra Leone when the democratically elected Kabbah government was toppled by

Local environment in rebel-hired PMSCs per year observation/major conflicts

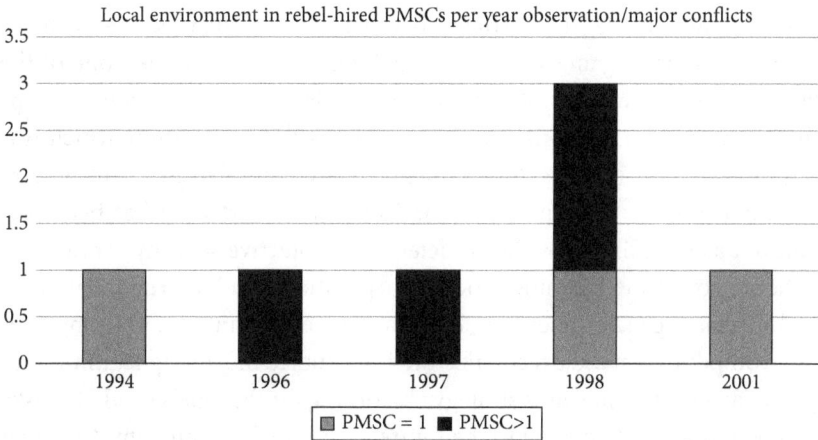

Figure 3.3  Local competition for rebel-hired PMSCs in major conflicts

Local environment for rebel-hired PMSCs per year observation/minor conflicts

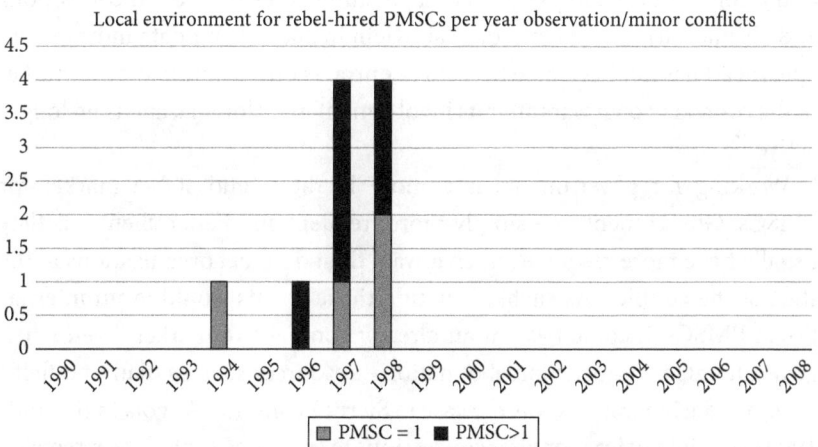

Figure 3.4  Local competition for rebel-hired PMSCs in minor conflicts

a coup in 1997. The PMSCs in this case were helping to bring back the previously ousted government that was democratically elected.

International PMSCs shy away from aiding rebel groups because working against internationally recognized governments could taint a company's reputation and be detrimental to its long-term profit-maximizing strategy. Contracting states are advised against working with actors that violate international norms. PMSCs that obtain a license from their home state for operating in conflict zones and violate it by assisting insurgents against state actors would suffer reputational costs. Most international PMSCs boast that they are respectable organizations and work only for established governments and legally recognized organizations such as private firms or international organizations. This distinction, in fact, is often one of the factors that separates PMSCs from mercenaries and ad hoc armed groups that are less selective in their choice of clients.[51] ArmorGroup, which was acquired by G4S in 2008, for example, clearly identified its client base to include legitimate governments: "for 25 years ArmorGroup has been recognized as a leading provider of defensive, protective security services for national governments, multinational corporations, and international peace and security agencies operating in hazardous environments."[52] Companies in a competitive market recognize the importance of gaining legitimacy as part of an effective business strategy.[53] Working for the rebels would be risky and potentially detrimental to the future survival of the company. Consider the case of Israeli private company Spearhead, which provided training to death squads hired by the Medellín drug cartel in Colombia. The revelation of the company's dealings by Colombian authorities resulted in convictions and ultimate arrest of its founder Yair Klein in 2007.[54] Our data indicate that Spearhead has not been able to secure contracts on the international market in the context of civil wars after its involvement in Colombia, and is no longer active.[55]

Working for governments is a more lucrative and stable market for PMSCs. Governments are simply more frequent and better clients, as they usually have more resources in civil wars than do rebel organizations at the start of the conflict. As such, alienating those clients would drain international PMSCs' income base in an already competitive market. Even when governments have lost control of national resources such as diamond fields or mines during wars, as was the case in Sierra Leone and Angola in the mid-1990s, the right to lease or offer concessions for resource exploitation remains in the government's domain. Francis (1999),[56] for example, alleges that when the government of Sierra Leone lacked the means to pay for the Executive Outcomes' military services that cost several million dollars a month, it offered the company commercial rights in the territory to be conquered. This was also the case for Sandline, whose contract in Sierra Leone was secured with a promise of mining proceeds.[57] In Angola, Executive Outcomes could count on access to profitable mineral rights that enabled the company to grow and diversify its services.[58] Not surprisingly, then, governments are in a better financial position to take advantage of PMSCs' services, making them the most lucrative clients. Consequently, we observe only a few cases of PMSCs assisting the rebels, and in those rare instances there are more cases of PMSCs operating as single providers than instances when local competition is present.

## Conclusion

This chapter explored the idea of local competition and examined its potential to increase accountability among PMSCs, with positive consequences for military effectiveness in civil wars. Unlike industry-wide rivalry, local competition reflects the number of providers that a client hires in a given conflict and exhibits greater levels of variation. There have been conflicts where governments have relied solely on the services of one company, while in other contexts numerous security providers contributed to the war effort. We argued that such variation in the number of providers is likely to affect companies' strategies in wars. As the number of providers increases, PMSCs are more likely to perform better because the client can compare their actions and quality of military effectiveness more directly than in instances where only one provider is present. One provider has the potential to exert greater leverage on the clients; it is, for example, in a position to manufacture security threats and make a case for contract expansion. By contrast, in contexts where local competition exists, a lackluster performance by one company could lead to its dismissal because the client can observe that another provider performed better. Considering that there are now many companies capable of offering multiple services in a wide range of areas—from guarding military convoys to providing logistical support—as a single package, it becomes easier for clients to replace a less-effective provider with a competitor.

Comparison is possible because rival companies serve as each other's watchdogs, informally monitoring what the other is doing to stay on top of the competition. This in turn is likely to push companies toward greater effectiveness in ways that would increase the odds of civil war termination. In the next chapter, we turn to the second dimension of market pressure, global competition, to examine how industry expansion has affected companies' strategies of growth. We explore how one particular strategy, shift in corporate structure, impacts companies' military effectiveness.

# 4

# Global Competition, Corporate Structure, and PMSCs' Accountability in Civil Wars

While local competition is one dimension of market pressure that PMSCs may face, the other is the general increase in the number of providers in the industry. The industry has expanded massively. In 1989, for example, there were 15 PMSCs on the market, while from 2003 to 2019 over 120 international PMSCs have been present in Iraq alone. Between 1994 and 2002, the US DoD awarded over 3,000 contracts to private contractors worth more than $300 billion.[1] Although not every company can rival established giants in the industry such as MPRI, DynCorp, Halliburton, or Kellogg, Brown and Root, companies have been taking notice of either new players entering the market or established ones expanding and diversifying their services.

Competitive pressure increases in an environment that has low barriers of entry for new firms.[2] This has been true in the case of PMSCs. In the late 1990s, men with military background have been able to start companies and secure lucrative contracts from the US government almost immediately after establishing their firms. Consider the case of Triple Canopy, an American company that won a contract to protect Coalition Provisional Authority (CPA) headquarters in Iraq. Founded by Matt Mann, retired US Army Special Operations master sergeant, and his friend Tom Katis, who was associated with Special Forces, Triple Canopy had limited experience with conducting operations in a conflict zone, including the particularly risky ones that involve escorting officials through dangerous roads.[3] Many retired military men with solid résumés could start a company and bid for a contract. In other cases, firms specializing in one service have become specialists in other areas after seeing the demand for private military and security contractors. In the midst of growing opportunities in Iraq, the British company Control Risks Group, for example, has expanded its repertoire of services after initially specializing in kidnapping negotiations.[4] The low barriers to entry for PMSCs have created an incentive for corporate security

*Private Militaries and the Security Industry in Civil Wars.* Seden Akcinaroglu and Elizabeth Radziszewski, Oxford University Press (2020). © Oxford University Press. DOI: 10.1093/oso/9780197520802.001.0001.

providers to offer potential clients a competitive edge in ways that could set them apart from their rivals.

Given that men with military experience have been able to start a private security business and successfully bid for contracts, it appears that clients have been less concerned about working with nascent companies. How, then, could companies respond to this new market pressure and gain an advantage over their competition? As argued, one of the key problems surrounding the phenomenon of outsourcing security to private companies is rooted in the principal-agent dynamic. Because PMSCs are profit-oriented businesses, their interests may differ from those of their client, thus creating problems with limited effectiveness. In essence, one issue that makes PMSCs consistently prone to criticism is the accountability deficit. Some have relied on this accountability factor as a sales pitch when bidding for a contract, highlighting, for example, their competitor's connection to human rights abuses and outlining positive changes in their own programs designed to deter such crimes in the future.[5] ArmorGroup North America has also boasted to US policymakers about being "first" in publishing the industry's white paper to regulate the industry.[6] The fact that companies are recognizing the need to improve their adherence to international humanitarian laws, in particular, is a result of extensive efforts by the international community to raise awareness about the issue of human rights violations in the private military and security industry.

Some companies have adapted to competitive pressure from industry-wide growth by transitioning from privately held to publicly traded corporations. The rise in the number of PMSCs going public especially after the year 2000 has come at a time when an increase in the quantity of firms but also in the quality and sophistication of services they are prepared to offer has emerged as a growing trend. A decision for a PMSC to go public is a strategic one and consistent with long-term expectation of maximizing profit. It enables the company to raise capital and use the money to invest in growth and the development of a unique competitive edge. Although not the norm, PMSCs also go public to signal their commitment to greater accountability as expected by mega clients, the US and UK. A notable case where accountability was cited as a factor is ArmorGroup North America. In a 2008 hearing before the Senate's Committee on Homeland Security and Governmental Affairs, James D. Schmitt, the company's senior vice president, mentioned that motivated by its commitment to transparency and accountability, the company became a publicly traded company.[7] PMSCs with public stocks increased in value at twice the rate of the Dow Jones Industrial Average in the 1990s.[8] For those companies that become public after being acquired by another public company, as was the case with MPRI when it was bought by L-3 Communications in 2000,[9] there is almost immediate access to the resources and networks of the parent corporation that often prove important in securing clients.

Media coverage of publicly traded companies is generally much greater than coverage of companies without a public stock offering, as more people have invested in corporations listed on the stock market. The public is more likely to attune itself to the news pertaining to publicly traded companies whose stock they may own. As the public's stake in companies rises, the media responds by paying more attention to such corporations.[10] This inevitably creates more publicity for the company, which in turn attracts attention from clients and thus increases the perception of the company's dominance in the industry.[11]

Furthermore, when a company goes public, its founders and board of directors often invest time and energy into developing a sound long-term business strategy. The process can be daunting; in fact, most publicly traded PMSCs have abandoned their private status years after being on the market. Yet the strategy of long-term business development connected with going public could credibly signal to potential clients the companies' awareness of industry woes as a whole, one of which is the need for greater accountability, and the recognition that media scrutiny comes with a change in corporate status. The idea, then, is that if an initial public offering (IPO) is years in the making, the companies must be serious about living up to the standards of being a publicly traded PMSC. Therefore, there is an expectation that such PMSCs should be more accountable to the clients and other stakeholders.

## Stakeholders, Commitment to Corporate Professionalism, and Military Effectiveness

While exposure to the media helps companies generate more publicity and visibility that could translate to interest from potential clients, it also contributes to making companies more committed to upholding corporate professionalism. A poor showing in the area of corporate professionalism, the dimension of military effectiveness that we highlighted in Chapter 2, that involves human rights abuses or fraudulent behavior such as deliberate manufacturing of security threats is not only much more likely to be published when committed by a publicly traded PMSC but also has the potential to engender greater reputational and financial costs than if similar publicity hit privately owned PMSCs. Put simply, given that publicly traded PMSCs have more stakeholders that they are responsible to, revelations of actions that undermine accountability to the client and generate negative reputation are especially likely to hurt such companies.

The media is one such stakeholder. Although it not a shareholder, it assumes the role of an indirect stakeholder because of its direct connection to the public, which is a shareholder and thus also a stakeholder.[12] From a client's perspective this heralds positive developments; it means that publicly traded PMSCs are likely to behave responsibly or else risk negative publicity. Criminal behavior, for example, an exposé of human rights abuses or revelations of fraud, is likely to garner more extensive coverage when committed by a publicly traded PMSC than by a private company because of the media's indirect stake in the company's reputation and performance.[13] Given that the public is not a stakeholder of private companies, the media's incentive to scrutinize the companies, while present, is lesser than for public companies.[14] Thus, the media's coverage of scandals may be less frequent in connection with private companies.

Second, the relevance of the media as an indirect stakeholder should be more significant in increasing accountability if the companies anticipate costs from bad publicity. Put simply, companies must believe that shareholders care about companies' reputation and thus might punish them for extensive media revelations of scandals. This is likely to be the case for companies that have more shareholders and thus more so again for publicly traded PMSCs than private ones. Chajet (1997),[15] for example, shows that the more shareholders a company has, the more likely it is that they will base their investment decisions not only on expected immediate value of the shares, but also on the company's reputation. The latter may affect profit in the long term but shareholders are not only concerned about those long-term financial gains. In fact, Chajet (1997)[16] argues that shareholders form their perceptions of the company's reputation through a complex process of social, collective, and psychological factors that are rarely limited to dividend returns. As the number of shareholders increases—a more likely scenario for public than private companies—there is greater likelihood that people want to be connected to "good" companies, ones that are not stained by scandals.[17]

Insights from the literature on corporate social responsibility, stakeholders, and reputation thus suggest that for publicly traded PMSCs, which have more shareholders and stakeholders than private PMSCs, the risk of incurring reputational costs due to bad publicity could be much higher. With more shareholders involved, initial loss of capital can easily lead other shareholders to lose trust in the company and generate a rapid downward spiral. This is likely to occur because shareholders form their views of the company's reputation based on affective cues that include social and collective psychology rather than on merely technical characteristics of a company. It is thus harder, though not impossible, for a publicly traded company to recover from such a loss once shareholders perceive the company as "shameful," a development that can be accelerated because of the media's unique position as indirect stakeholder.

DynCorp, which went public in 2006, attributed a massive drop in share prices to significant media coverage connecting the company to fraudulent practices in its training of the Afghan National Police in 2010–2011, coverage that the company has argued was much greater than that of other companies embroiled in similar situations.[18] According to Joseph Vafi, a stock analyst covering defense contractors, DynCorp "has been a tough public stock. A lot of what they do carries a lot of headline risk with it."[19] Although DynCorp eventually recovered from the scandals and continued to secure governmental contracts, the company recognized the inherent risk that comes with an increase in the number of shareholders and the inclusion of the media as a stakeholder. In 2008, for example, the company lowered its full-year profit outlook because it expected a delay in contracts due to the US Department of State's fraud investigations. In 2010, the company experienced a nearly 20% fall in its stock price in the midst of another fraud accusation.[20] These tangible costs and the possibility of more investigation of future operations were reasons for the company's decision to eventually abandon its public status in 2010.[21] This illustrates that publicly traded PMSCs are at a potentially greater risk of experiencing more scrutiny, and if they seek to maintain a strong profit margin they must either be more risk-averse with respect to violations of corporate professionalism or switch to private status. As private companies, they may be less likely to face contract delays or capital loss than public entities due to more limited scrutiny.

While publicly traded companies can recover from practices that could undermine military effectiveness, the risks associated with such a recovery both in terms of reputation and lost capital are much greater than for private companies. In an industry with low barriers to entry, they could be potentially devastating as substantial media investigation of possible mismanagement could result in the loss of a contract worth billions of dollars and spiral seemingly out of control. This is precisely what happened to DynCorp. With negative headlines damaging the company's reputation and shares tumbling, DynCorp had to drastically revise its strategy. Thus a deal was forged for Cerberus Capital Management, a private-equity firm, to acquire the company.[22] DynCorp, it seems, could not recover investors' confidence without a massive shift in its strategy. Contrast this development with the impact of the 1999 sex scandal in Bosnia on DynCorp's survival. Although the scandal, in which several DynCorp employees were accused of participating in a scheme to sell and sexually exploit women while engaging in peacekeeping operation,[23] tainted the company's image it had a less significant impact on the corporation's success. Not only was DynCorp able to recover, the company grew to become an industry giant. Thus its status as a privately held company in all likelihood made it easier for the company to recover faster than in the midst of the 2011 allegations, extensive media reporting, and subsequent decline in share prices that paved the way for Cerberus's acquisition.

This is not to say that privately held PMSCs would be immune to contract loss if fraudulent practices were uncovered; rather, the odds of uncovering such behavior are larger for publicly traded PMSCs because of greater media focus. Once the exposure is there and stock prices fall, future clients' confidence in the company could plummet, forcing the company to drastically revise its business strategy. Being a publicly traded company thus serves as a deterrent against low military effectiveness. When companies score high on corporate professionalism by recognizing the need to secure the trust of the locals and by adapting strict penalties against corruption at the top and low levels of the company, they are much more likely to help the government secure gains in a conflict zone and thus contribute to shorter wars.

It is also important to note that the risk connected to media visibility is further magnified for publicly traded PMSCs in light of the Security and Exchange Commission's requirement for public companies to disclose, among other things, outcomes of operations and any risks connected to company's investments, such as a lack of previous operating history or a proposal for highly speculative offerings.[24] While such practices may at first appear to shield companies from litigation, research shows that companies see mandatory disclosures as risky and connected to a spike in investor litigation.[25] Mandatory disclosures may create an opening for the companies to be more scrutinized, and the media can encourage such practices through its role as an indirect stakeholder. By priming investors to tune more closely to new information about behaviors that may be considered risky yet might have not been previously disclosed, the media provides investors with an opportunity to exercise more activism in punishing "bad" companies.[26]

If publicly traded PMSCs are more sensitive to reputational costs because they have more shareholders—the public that increasingly values companies' commitment to corporate professionalism—and the media that is an indirect stakeholder with an interest in revealing scandals as it serves the public, there is an expectation that such companies should be committed to taking corporate professionalism more seriously than private companies as a way to maximize access to lucrative contracts. Yet one of the challenges for the clients who value corporate professionalism and its contribution to military effectiveness is the ability to differentiate between those PMSCs that are credible in upholding a culture of integrity from those that are not. Incorporating corporate professionalism or corporate social responsibility into its business practice may not automatically suggest improvements in PMSCs' accountability. First, even though PMSCs heavily recruit from the military, there is no guarantee that companies' employees have stayed in the service long enough for the instillation of professional standards or that they will continue to comply with them outside the military.[27] Kinsey (2008) argues that PMSCs' acceptance of ethical norms then depends on the attitude of senior managers. Second, obtaining certifications on good practices as evident in the growing number of signatories to the International Code of Conduct for Private Security Service Providers shows greater initiative toward embracing corporate professionalism, but the challenge here lies in credibly communicating such a commitment in conflict zones where monitoring is daunting. In fact, Macleod and Dewinter-Schmitt (2019)[28] argue that internalization of corporate social responsibility among PMSCs, specifically practices involving respect for human rights, should come with greater oversight as such internalization takes time. Publicly traded PMSCs, however, can more credibly signal their commitment to corporate professionalism because of the large number of stakeholders that the company is accountable to.

In practice, publicly traded PMSCs' reputational sensitivity has positive consequences for improving military effectiveness. In the area of military and individual capabilities, it means that such companies are likely to invest in high-quality equipment and stricter rules for hiring qualified employees.[29] Programs are also designed to improve companies' adherence to international humanitarian laws and anti-fraud practices. For example, educating employees about interactions with the locals in conflict zones, better screening of employees, and emphasis on employee adaptability to complex conflict zones were among the initiatives that ArmorGroup pursued.[30] The company has adopted a policy of hiring only those employees with unblemished records, while others have implemented a zero-tolerance policy on abuses. After Pacific Architects and Engineers was acquired by publicly traded Lockheed Martin, on-field monitoring from the corporate office increased significantly. A general consensus emerged among the employees that corporate professionalism was taken more seriously than ever.[31] Some companies have focused on upholding humanitarian standards among its employees through initiatives designed to foster greater familiarity with indigenous cultures.[32] In order to minimize indiscriminate violence in a conflict zone and reduce civilian casualties, a company may also trade the use of powerful weapons for a more low-level approach.[33]

PMSCs that care about accountability to the client, meaning they seek to assist the government in ultimately securing a victory against an enemy, and that attach more value to their reputation because of market pressure would embrace a more active role in pushing for human rights inclusion in the contract. At the very least, they would be more likely to refrain from working for clients that push for strategies with the potential to backfire in terms of military effectiveness, such as mass killings of civilians, and that could severely damage the company's reputation both for failing to terminate the conflict and being linked to humanitarian abuses in front of potential clients. Avant (2005)[34] argues that because of their military expertise, PMSCs are in a position to be leading agents in defining and redefining threats and taking on a more dominant role in shaping contractual expectations, especially in weak states. In some instances, governments grant PMSCs a great deal of freedom to shape the contracts, as was the case with Executive Outcomes (EO). For example, EO came up with an idea of training Kamajors, the local hunters, to assist in its ground offensive against the rebels in Sierra Leone instead of relying on governmental troops for the task.[35] Thus, companies have opportunities to emphasize strategies that are consistent with corporate professionalism. These strategies need not mirror genuine ethical concerns but reflect a credible business model that aligns with the companies' profit-driven identity. This model recognizes that companies with a strong culture of integrity outperform other firms by a great margin in revenue growth, stock performance, and profitability.[36] For publicly traded PMSCs in particular it also reflects the awareness that such PMSCs face more severe reputational costs in a competitive market when ethical standards are neglected. Overall, then, we expect that publicly traded PMSCs should be more committed to upholding corporate professionalism than private PMSCs and thus should be linked to a lower number of human rights abuses and fraud in a conflict zone. A high level of corporate professionalism, in turn, should increase the companies' military effectiveness in civil wars.

It is worth nothing that the cumbersome strategies adopted by publicly traded PMSCs may be difficult to implement initially and temporarily offset the gains in a conflict zone, especially when they are novel. With time, however, the strategies are likely to increase the companies' accountability. When Lockheed Martin acquired Pacific Architects and Engineers, the rules changed drastically. The monitoring from the corporate office was on the rise and "everything was done by the book."[37] When at one point allegations of sexual harassment surfaced, the corporate office sent a team to pursue a lengthy investigation. Such practices were unheard of prior to PAE's acquisition by Lockheed Martin.[38] New regulations eschewed some difficulties for the teams on the ground. Adapting to new rules had temporarily lowered the employees' morale.[39] At the same time, the long-term benefits of corporate monitoring are apparent. With greater commitment to establishing a culture of integrity, the company signals its expectations for the conduct of its employees in the field. These expectations and the penalties for violations could deter employees from an indiscriminate use of force or from engaging in corrupt practices that with time could contribute to greater military effectiveness in a conflict zone.

### Publicly Traded Companies, Human Rights, and Fraud: Data and Insights

To examine whether publicly traded companies are more likely to embrace greater accountability, we collected data on alleged violations of two dimensions of corporate professionalism: commitment to anti-fraud practices and adherence to international humanitarian law. We investigate how the frequency of alleged scandals in the area of fraud and human rights abuses differs depending on PMSCs' corporate structure. We focus on Iraq (2003–2019), as the conflict there has received greater attention from the media and NGOs relative to other conflicts in which PMSCs have intervened, thereby increasing the reliability of data reports and reducing the problem of underreporting.[40] We examined reports of human rights abuses at the time of the companies' presence in the war. Our focus was on the abuses outlined in the fourth Geneva Convention on civilian protection, including arbitrary detentions and killings, torture, sexual crimes, and other instances of disproportionate use of force against civilians and property as well as discrimination. These are considered as worst-case scenarios and may not include low-level abuses such as aggressive posture or disrespectful behavior. A report that would fall under the category of humanitarian abuses might include a case of Blackwater killing a civilian bystander in the gunfire in 2005. Another example might involve accusations of abuse during interrogations of Iraqi youth by a British security company in 2004.

We also examined a second dimension linked to corporate professionalism, refraining from committing fraud. Here the focus is on noting fraudulent practices that involve evidence of deliberate intent to work against the clients' interest for personal gain. We include cases in which companies deliberately provide more and unnecessary services, engage in overbilling, take and give bribes, participate in criminal networks, pass on classified information to the enemy, sell illegal weapons, or illegally train a militia. A report showing that an American PMSC submitted false claims about unqualified security guards in Iraq would be an example of fraudulent activity.

We focus on PMSCs that are legally registered, international corporate entities with a clear business structure that deliver military and security services for monetary compensation.[41] The data consider abuses committed by a PMSC operating in Iraq as long as it is (a) a legally registered international corporation; and (b) has a business structure involving some form of managerial hierarchy, for example, including a CEO, vice president, etc. While indigenous security providers are on the market today, they are likely to face different competitive pressures from international PMSCs and thus merit a unique theoretical focus. We do not include them in the data. There are 122 PMSCs in our data, 29 of which are publicly traded.[42]

We relied on the following sources for our data: UN reports from the Working Group on the Use of Mercenaries; reports of abuses and mismanagement from Jurist, a legal news research service; evidence of PMSC misconduct tracked by the Business and Human Rights Resource Center; and evidence of waste, fraud, and abuse throughout the US government provided by the Project on Government Oversight (POGO). Additional reports came from independent organizations, including the Centre for Research on Globalization, Amnesty International, War on Want, Human Rights Corporate Watch, UNAMI, the Federal Contractor Misconduct Database, and Wikileaks. We also turned to LexisNexis, Google Books, online searches,[43] and *The Privatization of Warfare, Violence and Private Military & Security Companies*,[44] which compiles a list of PMSCs operating in Iraq and reports information about companies' services and human rights abuses.

While we recognize that some underreporting is possible, it is unlikely to be high and therefore have a significant impact on our inferences. Our data focus on the Iraq war, which has been extensively covered by the media. By relying on reports from NGOs with good knowledge of local developments, we can obtain information about developments from areas that might have been inaccessible to journalists. Finally, reporting cases of abuses and fraud is likely to be greater for Western companies or for companies that work for Western governments, as was seen with heavy media coverage of PMSCs' abuses in Iraq and Afghanistan. As most of the PMSCs in our data set are of Western origin, the problem with underreporting should be minimized.

The logic of our argument depends on the assumption that publicly traded companies are more likely to refrain from committing abuses than private companies because they are sensitive to media coverage of scandals. As such, continuous media coverage of the Iraq war is an important factor in the accountability story. While the coverage of private military contractors in Iraq has declined by half from 2011 until 2019 in comparison to earlier years, it remains fairly sizeable. This suggests that the media is still tuned in to contractors' presence in the conflict. We conducted a LexisNexis search using the following key terms: "Iraq" and "PMSC" or "private military company," "private security company," or "private military contractor," and the results show that from January 1, 2003, until January 1, 2011, there were 5,553 articles on the subject, while from January 2, 2011, until July 2019, 2,554 such articles were published.[45]

Another relevant issue concerns the legality of an activity and the standards by which it is judged in determining criminality. Standards are not always uniform. Drinking alcoholic beverages would be considered a criminal offense in Saudi Arabia but not in Western countries. We count an event as a human rights abuse or fraud when it violates the national criminal law of the states in which PMSCs operate. We also include all humanitarian abuses as defined by the Fourth Geneva Convention's section on the protection of civilians in armed conflict. States are expected to adhere to the standards of international humanitarian law, but in practice, it is sometimes the case that a violation by the international law standard might be ignored and no charges are brought against a PMSC. Even if the contracting state disregards a humanitarian violation, the nature of the violation does not change according to international humanitarian law and therefore would be included in our data set.

We also considered the murky distinction between failure to avoid a killing, a violation, and a necessary killing for self-defense. It is possible that some allegations of civilian killings might turn out to be false upon in-depth investigation, others deemed as acceptable acts of self-defense, while some confirmed as actual violations. PMSC contractors accused of killing civilians in Nisour Square argued that they began shooting after being fired at by Iraqi insurgents.[46] The case dragged on for eight years in courts, culminating in guilty sentences and ultimate dismissal of the self-defense argument.[47] Given that we include allegations and accusations of violations, it is possible that some percentage of allegations may fall into the murky category while some may be ruled as legitimate acts of self-defense upon conclusion of ongoing investigations. This represents a potential limitation of our data.

Figure 4.1 illustrates the frequency of human rights abuses and fraudulent practices based on the corporate structure of PMSCs that have been present in Iraq. Overall, the number of incidents involving human rights abuses is lower than that of fraud. There are 39 cases of human rights abuses committed by 18 out of 122 international PMSCs that have intervened in Iraq and 54 incidents of fraud committed by 24 PMSCs. Publicly traded companies are considerably less likely to commit human rights abuses and fraud than privately held PMSCs. Companies with private ownership have committed the majority of crimes in both categories: 34 instances of human rights abuses and 43 cases of fraud.

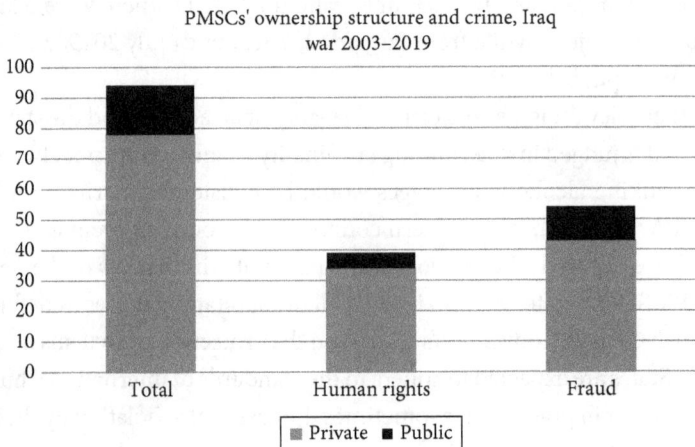

**Figure 4.1** Human Rights Abuses, Fraud, and PMSCs' Corporate Structure, Iraq War, 2003-2019

We conducted an empirical analysis to examine the impact of a company's business structure on its involvement in criminal practices. Our dependent variables, the number of human rights abuses and fraud committed by each company, are non-negative.[48] The data are cross-sectional; the entire duration of the company's presence in Iraq counts as one observation, unless the company became public during its involvement in the conflict, in which case we display two time periods to show the interruption in time. We have 137 observations, which is greater than the number of PMSCs that have intervened because sometimes companies change their status from private to public during the war, thereby increasing the number of observations.

It is possible that several other PMSCs' characteristics such as origin, age, size, structure of subcontracting chain, and membership in trade associations could also explain variation in the extent to which companies are connected to fraud and human rights abuses. In light of the scandals associated with the American PMSC Blackwater, American or European companies might be more risk-averse when it comes to scandals because of greater media scrutiny and openness. Larger and older companies can boast greater advantage in adaptability, resources, and experience necessary to invest in employee training in the area of professionalism, less lethal military equipment and innovative field training to reduce indiscriminate violence, as well as more investment in anti-fraud policies.

We consider the structure of the subcontracting chain, which could adversely affect monitoring. Principal-agent logic might suggest that longer chains of subcontracting might lead to greater deviation from the principal's requests because it might be harder for the principal to detect less accountable behavior when services are not under direct discretion of the prime contractor.[49] We thus examined whether companies have subcontracted their services to others.

Whether a company is a member of ISOA (International Stability Operations Association), BAPSC (British Association of Private Security Companies), SCEG (Security in Complex Environments Group), or ICoCA (International Code of Conduct Association) could indicate an interest in improving its approach to accountable practices that could potentially translate into more ethical behavior. Consequently, we also noted companies' membership in such associations, with the expectation that PMSCs with a larger number of memberships might be more serious about embracing accountability.[50]

Factors related to the client, specifically whether the client is the US government, another government, international organizations, NGOs, or other companies, might affect PMSCs' propensity toward criminal behavior. Most of the companies in Iraq have worked for the US government (78%). Nearly 60% of the companies have performed services for non-state actors, while 16% of PMSCs have worked for other governments. Our expectation is that governments may be better at monitoring company performance than non-state actors due to the greater number of resources that agencies could devote to oversight and policy development.

Certain types of services create more opportunities for greater contact with local population or are large enough to encourage overbilling. For example, companies delivering security services such as the provision of guards or convoy security may be more likely to engage in combat and accidental shooting that could lead to the killing of civilians. This could also be true for the training of Iraqi police and military personnel. Intelligence services may be more linked to human rights abuses, especially during human interrogations as was evident with the Abu Ghraib prison scandal involving US military and private contractors torturing Iraqi prisoners.[51] Services in construction, base support, and logistics could be linked to greater frequency of fraud as the complexity and the higher pricing of these services make it easier to overbill while hiding actual expenses.

Lastly, it is possible that earlier stages of the war in Iraq might have created greater opportunity for companies to engage in human rights abuses and fraud. We argue that competitive pressure has pushed companies to develop interest in corporate professionalism as a way to differentiate themselves from other security providers. Yet in the earlier stages of the Iraq war, that market was less competitive and thus companies', DoD's, and other agencies' consideration for ethical behavior, while present, might not have been systematically applied. The number of available jobs was high, leading to the rise of the so-called Baghdad Bubble.[52] According to Lawrence Peter, the private security assistant in Iraq, "if you wanted a contract, you were going to get a contract."[53]

Prior to 2008 there were also notable issues with contracts that might have encourage fraudulent practices such as overbilling and hamper incentives to develop approaches to limit human rights abuses. Contracts, for example, were vague and limited in setting both human rights standards and benchmarks for performance.[54] It was also more common to see the prevalence of indefinite delivery/indefinite quantity (IDIQ) contracts during the early stages of the war. Such contracts, which ensure that clients can receive services on a short notice over a fixed period, may create more opportunities for fraud. In a hearing before the Committee on Homeland Security and Governmental Affairs in 2006, Senator Carl Levin noted the possibility that an "IDIQ contract lends itself to abuse because when we finally decide what work we want done, when we do that, we will have no competition. As a result, we pretty much have to take whatever estimate the contractor offers."[55]

In the later stages of the war, competition has increased as more companies have diversified their service offerings. Revelations of waste and human rights abuses by the media in the wake of the Blackwater shootings, together with the Montreux Document, which was published in 2008, have put pressure on the US government to improve accountability and contractual expectations. This focus on emphasizing human rights protection and developing standards for PMSCs did not begin with Montreux. The DoD, for example, was aware that it needed high-quality management standards unique to PMSCs as early as 2004,[56] but Montreux is regarded as the cornerstone in clarifying international obligations[57] and an "entirely new framework for governing the industry."[58] As such, 2008, the date when Montreux was published, marked the time when normative guidelines were established for states that use PMSCs. Subsequent efforts such as ICoC (2010) and ANSI/ASIS Standards (2012) built upon it, but the Montreux Document was the initial step in formally establishing good practices for working with PMSCs. Thus, it could be considered the turning point for improving accountability at the international level. To account for the possibility that less competition in the early years of the war, coupled with contract vagueness, non-competitive contract renewal, dominance of IDIQ contracts, and limited international focus on guidelines for good practices, could have adversely affected companies' incentive to behave more ethically prior to 2008, we control for the first year in which a company began to serve in Iraq.

Tables 4.1 and 4.2 show the results for human rights abuses and fraud. We present findings from three models; each model includes our key explanatory variable (publicly traded) with different sets of control variables.[59] Results show that publicly traded companies are less likely to commit fraud and human rights abuses than private companies across all models. The models demonstrate that a PMSC's corporate structure is a credible deterrent against human rights abuses and fraud committed in a conflict zone. While both private and publicly traded companies have to some extent recognized the importance of investing in corporate professionalism, our findings show

Table 4.1  PMSCs' ownership structure and human rights: empirical results

| Human Rights | Model 1 | Model 2 | Model 3 |
| --- | --- | --- | --- |
| Size | 0.98(0.41)** | 0.54(0.36) | 0.68(0.40)* |
| Publicly Traded | −1.55(0.90)* | −2.20(1.02)** | −1.60(0.95)* |
| American | −1.10(0.69) | 0.28(0.57) | −0.38(0.64) |
| European | −1.61(0.78)** | −1.51(0.76)** | −1.14(0.77) |
| Company Age | −0.03(0.02)* | −0.05(0.03) | −0.04(0.02) |
| Subcontracting | 3.71(1.24)*** | 1.39(0.62)** | 1.83(0.76)** |
| Associations Membership | −0.89 (0.44)** | | −0.02(0.33) |
| Intelligence/Risk Assessment | 1.88(0.66)*** | | |
| Training | −0.11(0.52) | | |
| Demining | 1.73(0.68)** | | |
| Security | 2.93(0.90)*** | | |
| Logistics/Base Support | −0.53(0.73) | | |
| Other Services | −1.09(0.80) | | |
| Hired by US | | 3.02(0.72)*** | |
| Hired by Other Governments | | 2.48(0.75)*** | |
| Hired by Organizations/ Companies | | 1.15(0.53)** | |
| Less Competitive & Montreux | | | −0.88(0.61) |
| Constant | −4.44(1.18)*** | −4.70(1.14)*** | −1.48(0.75)** |

N: 137, *p<0.1, **p<0.05, ***p<0.01, reference for "company origin" is other than European and American

that when companies boast of initiatives such as a zero-tolerance policy for employees accused of abuses, employee training on cultural sensitivities, emphasis on low-level weapons approach to limit indiscriminate violence, and measures to curb fraud, such initiatives are indeed more credible when implemented by publicly traded companies. It is important to note that this relationship holds even when we take into account the difference between Iraq's less and more competitive environment and the pre- and post-Montreux era (Table 4.1 and 4.2, Model 1). Our results show that these developments on their own have been insufficient in deterring criminal behavior, as the variable is not statistically significant. It is likely that companies have approached demands from the US government to improve

Table 4.2  PMSCs' ownership structure and fraud: empirical results

| Fraud | Model 1 | Model 2 | Model 3 |
|---|---|---|---|
| Size | 0.68(0.23)*** | 0.51(0.22)** | 0.71(0.24)*** |
| Publicly Traded | −1.38(0.47)*** | −1.07(0.58)* | −1.32(0.59)** |
| American | −0.98(0.62) | −0.30(0.49) | −0.42(0.48) |
| European | −2.72(1.02)*** | −2.88(1.06)*** | −2.66(1.00)*** |
| Company Age | −0.02(0.01)** | −0.01(0.01) | −0.01(0.01) |
| Subcontracting | 1.63(0.59)*** | 0.87(0.58) | 1.28(0.58)** |
| Associations Membership | −0.81(0.35)** | | −0.47(0.36) |
| Intelligence/Risk Assessment | 1.41(0.74)* | | |
| Training | −1.20(0.59)** | | |
| Demining | 0.90(0.68) | | |
| Security | 0.25(0.50) | | |
| Logistics/Base Support | 1.31(0.53)** | | |
| Other Services | −0.13(0.50) | | |
| Hired by US | | 1.34(0.66)** | |
| Hired by Other Governments | | 1.15(0.66)* | |
| Hired by Organizations/ Companies | | −0.20(0.54) | |
| Less Competitive & Montreux | | | 0.07(0.38) |
| Constant | −1.74(0.94)* | −2.76(0.81)*** | −1.85(0.76)** |

N: 137, *p<0.1, **p<0.05, ***p<0.01, reference for "company origin" is other than European and American

accountability differently, with some companies continuing to take risks while others bolstering their existing efforts to improve in this area. Publicly traded companies might have already been less inclined to engage in criminal behavior than private companies but might have invested more in ethical practices in the post-2008 milieu to further minimize risk of abuses due to their greater sensitivity to stakeholders' interests. That might have been less of a case for private companies despite growing competition and more specific contractual expectations designed to limit abuses in the later stages of war. Therefore, when considering these two findings on corporate structure and the opportunity for abuses, our results show that the unique vulnerabilities of publicly traded PMSCs are a stronger predictor of companies' ethical behavior, and it is likely that more competition and client demand for greater accountability could push publicly traded PMSCs to become even more risk-averse than is the case for private companies.

In addition to our key findings regarding a company's structure, there are other factors that help explain variation in the frequency of crimes. Larger companies are more likely to engage in fraud; this relationship holds across all models (Table 4.2, Models 1–3). This suggests that monitoring problems due to possible struggles with coordination and task delegation may abound in larger companies and encourage less ethical behavior. In the area of human rights abuses, larger companies are also more likely to engage in such criminal practices, though the result is not statistically significant when the nature of hiring client is taken into account (Table 4.1, Models 1–3). Older companies are less likely to engage in criminal behavior (Table 4.1 and 4.2, Model 1); however, age becomes less relevant when we control for companies' membership in trade associations, hiring client's identity, subcontracting, and the pre-competition/Montreux milieu (Table 4.1 and 4.2, Models 2 and 3). European companies, overall, fare better in monitoring the actions of their employees with both types of crimes. The variable only loses statistical significance after controlling for membership in trade organizations and the pre-competition/Montreux time period for human rights abuses. Companies that have worked for the US government, other governments, international organizations, NGOs, or other companies are more likely to commit human rights abuses, indicating that the US is not unique in its struggles with oversight (Table 4.1 and Table 4.2, Model 2). Governments struggle more than international organizations, NGOs, or other companies in limiting PMSCs' from committing fraud possibly because these actors are more likely to hire for large and complex jobs where performance is harder to track.

Turning to subcontracting, our results show that companies with a subcontracting chain are more likely to be linked to both types of crimes because delegation to other companies brings its own agency problems (Table 4.1, Models 1–3, Table 4.2, Models 1 and 3). The fact that subcontracting is a bigger problem in the area of human rights abuses than fraud is likely because subcontracting firms usually hire smaller and more specialized companies that might not have devoted efforts to training their employees on humanitarian practices and are already at a higher risk of failure simply by being in direct contact with the locals more frequently.

The type of service that companies perform could also affect the extent of crimes they commit. Accountability in the area of human rights is lower for companies that deliver intelligence, engage in demining, and provide security. Greater contact with local population in provision of such services increases the odds of humanitarian abuses (Table 4.1, Model 1). Intelligence, logistics, and base support are tied to greater instances of fraud, while training is connected to limited cases of fraud (Table 4.2, Model 1). It is likely that the former types of services are linked to contracts that are bigger in size and thus create more opportunities for understaffing that might go undetected. Companies might be aware of limited oversight in such contexts and thus respond by taking more risk with overbilling the client. Finally, membership in trade organizations serves as a deterrent against both fraud and human rights abuses. However, the variable is not robust and loses its significance once we take into account the pre-competition/Montreux era.

Overall, our findings suggest that PMSCs' characteristics in terms of business structure, size, age, membership in trade organizations, subcontracting practices, client identity, and origin shape the companies' propensity toward human rights abuses and fraud. The most surprising finding is the lack of statistical significance for the pre-competitive/Montreux milieu in affecting companies' behavior. While these findings do not diminish the value of greater pressure that has been put on companies in the later stage of the war due to more competition and initiatives to encourage good practices, they suggest that what might matter more is how companies perceive the meaning of change as it relates to the opportunity to commit abuses. Whether that change is internalized might be more affected by the perceptions of hiring clients' monitoring capacity as well as companies' varying reputational costs and sensitivities to perceived changes in the accountability landscape. It is entirely possible that both public and private PMSCs have responded to this changing environment by claiming they embrace corporate professionalism, yet publicly traded companies, which are the minority of all international PMSCs, might have nevertheless exhibited greater commitment to such practices than private ones. This might explain why overall we do not see any statistically significant difference in the more competitive, post-Montreux stage of the war.

When we consider variables that we included in all three models and that are robust across all, there is strong evidence that a publicly traded status of a PMSC is a strong predictor of a company's decision to credibly commit to corporate professionalism. It is important to note that since deterrence is not perfect regardless of the different characteristics of PMSCs and services they provide, there is room for initiatives to enhance monitoring. For example, clients should be concerned about companies that subcontract, and thus to minimize the risk of subcontracting and human rights abuses, clients should demand stronger evidence of how companies vet subcontractors. Overall, it is vital to recognize that PMSCs' performance in conflict zones likely comes with potentially higher risks of abuses than engagements in other contexts due to higher levels of uncertainty inherently linked to war zones. Thus, success in conflicts might be measured with incremental improvements in companies' commitment to ethical standards rather than aiming for an all-out elimination of scandals.

## Publicly Traded PMSCs and Civil Wars: Trends & Implications

Our data indicate a growing presence of publicly traded PMSCs in civil wars until 2008. Figure 4.2 shows that before the year 2000 there was only one publicly traded company operating in one conflict, the civil war in the Democratic Republic of the Congo. Starting with the year 2000, the number of minor conflicts where publicly traded PMSCs operate had risen to nine. In the year 2000 one publicly owned company, MPRI, a subsidiary of L-3 Communications, fought on behalf of the Colombian government against the insurgents. The company helped armed forces and the national police with psychological operations, training, logistics, intelligence, and personnel management.[60] In the year 2008, there were 9 conflicts in which publicly

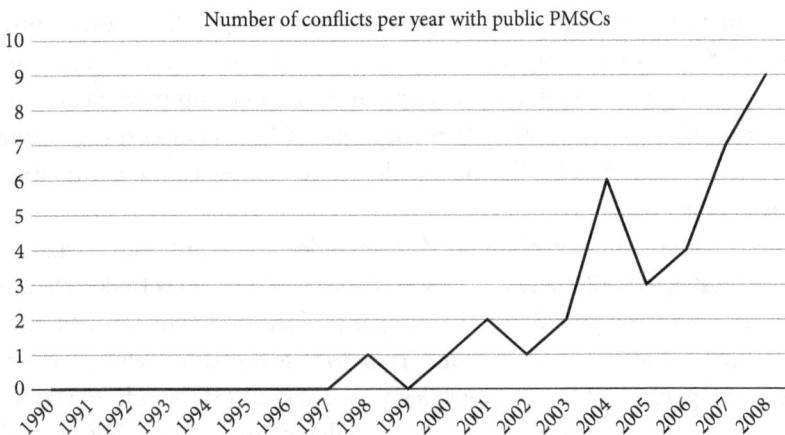

Figure 4.2 The presence of publicly traded PMSCs in minor civil wars

traded PMSCs were serving governments simultaneously outside of Iraq and Afghanistan, war zones where the presence of publicly traded companies was considerably higher than in other conflicts. Publicly owned PMSCs have been present in conflict zones across different regions, for example, Uganda and Nigeria in Africa, Georgia in the Caucasus, Israel and Iraq in the Middle East, Macedonia in Europe, Colombia and Peru in Latin America, and Pakistan and Afghanistan in South-Central Asia. Furthermore, our data also reveal that only government-hired PMSCs are publicly traded, while all of the rebel-hired PMSCs are privately held. This is expected, given that publicly traded companies in particular face greater media scrutiny and audience costs for bad practices. It is likely that if their contracts with the rebels were exposed, the reputational costs would be devastating to the company's long-term survival.

Overall, more publicly traded companies have worked in minor conflicts, as PMSCs have recognized the value in modifying their corporate structure and have thus adapted accordingly to the growing competition at the industry level. There are no publicly traded PMSCs that have intervened in major conflicts in our data set. Iraq, which we analyzed here, is the only exception. Companies began to change their corporate structure due to a rise in global competitive pressure, which has intensified considerably in the early 2000s. Prior to that, the industry, while competitive, was not saturated with so many new players. As such, the push for PMSCs to alter their business strategies and obtain a competitive edge might have been less pronounced. Yet it is after the year 2000 that there is a substantial decline in major conflicts; instead, minor conflicts begin to dominate in frequency. Consequently, the rise of publicly traded PMSCs coincides with the proliferation of minor civil wars and the decline of major ones. In the next chapter, we examine how the presence of publicly traded PMSCs has affected civil war dynamics. We explore whether working with more accountable publicly traded PMSCs is beneficial to the termination of conflicts.

For all the benefits that come from hiring publicly traded PMSCs that we highlighted in this chapter, it is nevertheless worth considering whether a client would still choose to work with a reputable company if it had an option of working with one that was offering more affordable terms but carried a murky reputation. Is the cost more important to clients than the level of accountability that a company can offer? The logic of our argument hinges on the idea that clients interested in working with international PMSCs care about reputation and reward those companies that demonstrate accountable behavior. Those interests may differ for clients who seek the assistance of mercenaries and ad-hoc companies, a market for force that extends beyond our core argument. In the former context, however, it is likely that the clients' interest in winning the war will converge with their interest to choose more accountable companies. Even when some clients within the PMSC market care less about humanitarianism, they still care about results. As we argued in Chapter 2, we expect companies with greater corporate professionalism to deliver more-effective strategies, and the flexibility in strategy design creates an opening for PMSCs to push for those that align more closely with international norms.

In the context of choosing price over reputation, the possibility exists that some clients may initially favor the former over the latter. Still, over time, we expect a shift in such interests once clients recognize the limited benefits of working with less reputable PMSCs. This mirrors the overall changes in consumers' evolving expectations regarding their purchasing choices in general and their support for corporate values more specifically. The 2017 Cone Communications Corporate Social Responsibility study found that 75% of consumers would refuse to purchase a product if a company supported an issue contrary to their beliefs. These findings are not merely limited to attitudes but are reflected in actions. The same study found that over 50% of respondents boycotted a company for "irresponsible business actions."[61] In the world of clients of the private military and security industry, "irresponsible business actions" is equivalent to low accountability or hiring companies that are expected to offer limited effectiveness in helping the governments secure a quick victory. Such companies might be staffed by inexperienced individuals, use the same people in day and night shifts to cut on costs, or engage in overbilling that wastes the scarce resources of the client and endangers the mission. It may be rational for a client to choose a low-cost company that is new on the market and for which there is no information on behavior in conflict zones. It is only with time and information that clients will be able to make more rational calculations between low cost/weak accountability and high cost/high accountability. It is then that clients are likely to choose the more accountable company over the one with questionable practices.

Companies that offer low-cost services yet have a reputation for limited interest in upholding corporate professionalism are also more likely to perish in the international PMSC market unless they rebrand or choose to work for clients that outsource to mercenaries and whose interests in conflicts are more diverse. The case of Crescent Security, a Kuwaiti company, illustrates the value of accountability over cost. Crescent Security was on the market for only three years when it was awarded a contract to protect convoys in Iraq. While not much was known about the company at the time, Crescent Security offered affordable services at the expense of forgoing important safety regulation and violating US military regulations by purchasing illegal weapons. Upon discovery of unscrupulous behavior, the US banned the company from operating in Iraq. Unable to rebrand and secure new contracts despite the low prices for its services, Crescent Security soon went defunct.[62] Overall, while a few resource-deprived clients might resort to working with companies whose reputation is poor but which offer low-cost services, many would likely opt not to work with anyone rather than take the risk of wasting valuable resources on securing providers that are unlikely to be very effective in contributing to conflict termination.

## The Market's Limitations

Recent research on PMSCs and the markets in which they operate argues that the global market for force is not singular, but rather it consists of three distinct markets that have divergent consequences on states' control over force and the provision of security as a public good.[63] Furthermore, the structures of some of these markets, at times, do not operate according to the neoliberal market dynamics we described in this chapter. In a neoliberal market, private companies compete with each other for a contract based on quality of service and cost. In the world of PMSCs, this is the most dominant market, found in the US, Europe, parts of Latin America, and Asia. Our argument about the impact of local and global competition applies to this international, neoliberal market.

Yet other types of markets exist. In China, a hybrid market prevails, in which the government both controls and privatizes security companies. Until 2010, when the Chinese State Council pushed forward the Security Service Management Regulation Act with the goal of opening the market for foreign security companies, the market made it difficult for foreign companies to openly enter and compete for contracts.[64] Such preferential treatment for domestic security companies means that market pressure stemming from global expansion of the industry is less relevant in a hybrid context. While Chinese companies still compete with each other for domestic contracts, that competition has, until 2010, been restricted to domestic actors. By contrast,

PMSCs competing in the international market face a more competitive environment characterized by an influx of new players on a yearly basis.

The impact of neoliberal market forces on PMSCs' accountability will also be limited in racketeer markets where criminal organizations engage in intimidations rather than in legal means of gaining a competitive advantage to secure a contract.[65] In such contexts, private security providers do not compete based on standards developed by the international community on the hiring of PMSCs and objective measures of military effectiveness but turn to force and intimidation to exclude the competition. The consequence here is that the state loses its monopoly on force in such markets.

There is also the question of oligopoly in the private security market, especially in Iraq, Afghanistan, and Colombia, where the trend has been for the US to reward the same large companies, such as DynCorp, MPRI, KBR, and Vinnell, with contracts, a phenomenon that leads to unfair competition and undermines the market's regulatory effect. Between 2003 and 2008, 75% of all US contracts in Iraq went to ten major PMSCs.[66] The industry giants may be ahead of smaller companies in these three markets, but they still face some level of competitive pressure from each other and the possibility of being monitored. Consequently, revelations of opportunistic behavior by whistleblowers could result in a rival giant getting the contract. The market in such contexts is still competitive, although the competition is indeed restricted to the bigger companies. A similar issue emerges with mergers and acquisitions that push smaller companies out of the market. Between 2002 and 2008, for example, G4, the largest private security company, acquired Wackenut, Securior, and the widely known British PMSC ArmorGroup.[67] The ability of one company to offer a myriad of services creates an incentive for clients to hire only one player, thereby restricting competition. This could limit the impact of local competition as a mechanism of accountability that we introduced here. Yet even here, the competitive pressure does not completely disappear. Big companies compete against each other and thus face constant pressure to secure a competitive edge.

The market's impact on shaping the behavior of PMSCs might also be curtailed when the market merely appears to be competitive due to the existence of many security providers, yet the abundance of opportunities for contracts makes it, in reality, less competitive. In such contexts, although multiple providers might operate in a conflict zone, they may have less interest in monitoring each other and reporting human rights abuses and fraud. This was the case, for example, in the early years of the Iraq war. In

Iraq, the local environment appeared competitive as private contractors performed numerous jobs from delivering food and supplies to the troops to assisting troops in insurgent patrols and guarding diplomats. Yet the competitive pressure was less significant in Iraq in the early stages of the war because of the large number of available jobs,[68] and the presence of two clients, the US government and the Iraqi government. Because the demand was large, former military men could easily establish a company and get a contract from the DoD. As such, we would argue that despite the presence of many PMSCs in Iraq, there was less incentive for companies to ponder the rivals' performance and monitor their behavior. The pressure, however, grew after the Blackwater 2007 shooting scandal. After the incident, the US government limited the number of contractors across the industry and opportunities began to diminish.[69] In the aftermath of the scandal, PMSCs begin to improve their adherence to international humanitarian laws in part also due to better monitoring from the US government.[70] This is evident in the decrease of crimes committed after 2009.[71] Cases like the Iraq war, however, are an outlier. Our data set shows that abundant contracting opportunities in civil wars are unusual mostly because the majority of contracting clients in civil wars are not the US. These clients lack the resources to offer numerous opportunities to contractors. Besides Iraq, other contexts in which we see two contracting states in the same civil war and thus the possibility of somewhat greater contract opportunities are Afghanistan, Colombia, and Peru. In all other cases, only one state issues the contract in a given conflict. When we consider the trend of a single contracting state and the limited resources of most contracting states in civil wars, we can conclude that opportunities for contractors in a given conflict are not plentiful enough to decrease competitive pressure among these actors.

Subcontracting of services by prime contracting PMSCs is another challenge to the accountability model. Prime contracting companies receive lucrative contracts and hire smaller and, at times, more specialized companies to fulfill parts of the contract. This presents challenges in maintaining accountability when military and security services are not under the direct discretion of the prime contractor.[72] Our analysis, in fact, shows that companies with a subcontracting chain are more likely to be linked to abuses. Still, publicly traded companies that care about their reputation are more likely to include stricter, zero-tolerance policies from the start for subcontractors in the same way they do for smaller companies they acquire. This was the case with Pacific Architects and Engineers employees, for example, who were exposed to more robust monitoring and faced a stronger likelihood of dismissal for violating the rules after Lockheed Martin acquired the company in 2006.[73] Such companies can communicate zero-tolerance policies not only to their own employees but to subcontractors as well and choose to work only with more reputable companies. Given that co-nationals carry the most difficult jobs in conflict zones in terms of strategy and risk,[74] there are more opportunities to vet such employees. The problem would be greater if subcontractors worked mostly with third-party nationals, who might be harder to vet, but that is not the case for challenging jobs. Overall, the pressure to vet subcontractors has increased over time. The National Defense Authorization Act of 2008 and 2011 requires that companies vet their subcontractors and that such a clause is included in contracts coming from the US DoD.[75] Companies acknowledge their responsibility for subcontractors. The same regulations are also found in the ANSI/ASIS Standard and ISO Standard and thus apply to companies operating internationally. Therefore even though the problem with subcontracting is greater when the prime company uses subcontractors who, in turn, subcontract, with the chain of subcontracting getting longer, more emphasis on vetting could improve accountability, with the expectation that such pressure would be even greater for publicly traded companies due to higher reputational costs.

A market approach cannot succeed unless the clients are willing to hold the companies accountable for weak military effectiveness. In most circumstances, we might expect the client to do so, yet the state's bureaucracy might at times impede the efforts to reward good practices. In the early years of the Iraq war, absence of strong monitoring incentives among PMSCs, coupled with US governmental agencies' limited concern for evaluating performance and their contradictory approach toward managing private contractors, had a detrimental effect on accountability. Loyalty, personal networks, and preference for hiring American companies were among the factors affecting selection of some PMSCs over others, demonstrating the consumer's reluctance to exercise effectiveness as a criterion on contract awards.[76] The US as a mega consumer failed to exercise its strong market power to set standards for performance as it continued to renew contracts for companies with limited regard for accountability.[77] CACI, for example, was awarded another contract in 2005, a year after its reported abuses at Abu Ghraib prison. Similarly, Blackwater's contract was extended for another year, despite the Nisour Square killings.[78] Yet the market eventually ends up punishing those companies especially when democratic clients, such as the US, face growing pressure from the media, the public, and the international community to uphold higher standards. Blackwater, for example, was subjected to an FBI investigation, termination of contracts by various departments, and prosecutions by the Justice Department.[79] While it is up to the client to exercise its power to reward or replace companies based on their effectiveness, domestic and international pressure can help push the client to eventually penalize less effective companies.

Finally, competitive pressure might tempt some companies to engage in defamation, a phenomenon that could distort the market's effectiveness in rewarding companies with good practices and undermine fair competition. Lengthy court cases are expensive and generate negative publicity that might divert companies' resources from growth, innovation, and investment in cultivating a culture of professionalism that benefits many of their clients and contributes to a more accountable security market. Even more damaging, news of accusations could stain a company's reputation and the devaluation of its stock if a company, is publicly traded. Yet there is little evidence that defamation has become a frequent by-product of informal peer monitoring. Accusations of misdemeanor by a rival company must still be investigated and if a rival company engages in habitual defamation with limited evidence, for example, lacks credible eyewitnesses, photographs, emails, or recordings to document an abuse, it is likely to inflict reputational costs on itself for engaging in unprofessional behavior. In the context where legal mechanisms are weak, the target can still provide counterevidence to the client or the media to save its reputation and discredit the accuser. If a company habitually engages in defamation and such accusations are found to be false or weak, then it risks damaging its own reputation.

Additionally, defamation brought by an individual employee against a company is also costly to that employee, therefore reducing the likelihood of such incidents. In the case of *Mayberry v. Custer Battles*, the judge called for an investigation into perjury against Rory Mayberry for false accusations against a PMSC.[80] Given that international PMSCs are increasingly engaging in more in-depth screening of their personnel, legal investigation of deception could carry a long-term negative consequence for employees seeking to exploit the monitoring phenomenon for personal gain.

In instances when defamation involves only non-Western international PMSCs delivering services in weak states, there is a greater potential that actors might not receive proper legal consideration or media attention to make such behavior risky. If international PMSCs are registered in states with corrupt legal systems and an inadequate free press, they face the possibility of corrupt judges and inadequate attention to evidence. It is in those cases that accusers could potentially gain an unfair competitive advantage because the rival company's reputation could now be stained. Here, the market could fail to punish opportunistic companies.

## Conclusion

Scholars have already recognized that low barriers to market entry have contributed to an unprecedented expansion of the private military and security industry, with competition becoming increasingly fierce. Our argument builds on this idea by showing that in response to competitive pressure some PMSCs have strived to develop a competitive edge that could separate them from their rivals and make them attractive to potential clients. Given that many companies on the market have been around for a long time and offer sophisticated and highly diversified services, some PMSCs have focused on gaining capital and signaling to future clients their commitment to accountability to differentiate themselves from the rest. There is strong consensus among the industry's critics that limited accountability has a detrimental effect on PMSCs' behavior in conflict zones. In an attempt to make profit and reduce costs, PMSCs could be tempted to obtain quick though fleeting gains in a conflict by abusing the local population. From a client's perspective, human rights abuses committed by PMSCs could hurt the government's ability to terminate the war, as would fraudulent practices and reliance on inadequate material capabilities.

To primarily gain more capital but also to credibly communicate greater accountability that would mitigate such scenarios, some PMSCs have opted for more openness by becoming publicly traded companies. Because they face greater media scrutiny due to their publicly traded status and risk significantly greater attention from the media and the public in the midst of scandals, such PMSCs are much more likely to foster corporate professionalism and invest in better-quality equipment and employees. Empirical analysis of PMSCs that have intervened during the Iraq war shows that a company's publicly traded status serves as a deterrent against committing human rights abuses and fraud. Greater sensitivity toward human rights and anti-fraud practices should improve PMSCs' military effectiveness. In the next chapter, we examine empirically the validity of our argument by looking at the impact of PMSCs' interventions into major and minor civil wars under varying levels of global and local market competition.

## Appendix 4.1   PMSCs' Interventions in Iraq, 2003–2019

---

**List of PMSCs Providing Services in Iraq, 2003–2019***

---

Aegis Defense Services Ltd.
Agility Logistics (formerly PWC)
AKE Group
Alfagates
Allied International Consultants and Services, AICS
Amarante International
Anticip
Ardan Consulting
ArmorGroup International plc
Asbeck Armoring
Babylon Gates
Bearing Point
BH Defense
Black Hawk Security
Blackwater
BLP
Blue Hackle
Britam Defence
CACI
Castleforce
Castlegate-CSS-Iraq
Centurion Risk Assessment Services Ltd.
Cochise Consultancy, Inc.
Computer Science Corporation (CSC)
Control Risks Group
Covenant Homeland Security Solutions Inc.
Crescent Security Group
CTU Security Consulting Inc.
Cubic Corporation
Custer Battles LLC
Danubia Global Inc.
Diligence LLC
Dimensions International, Inc.
Double Eagle MC—Tactical & Material Support Services
DTS Security LLC
DynCorp International
ECCI, Environmental Chemical Corporation International
Edinburgh International
EHC Group
EODT (also known as EOD Technology, Inc.)
Eryns International Ltd.
G4S
Gallice Security
Garda World
Genric Security

## List of PMSCs Providing Services in Iraq, 2003–2019*

Global Strategies Group (Global Risk Strategies)
Globe Risk International
Greystone
Groupe GEOS
H3 High Security Solutions LLC
Hart Security
Henderson Risk Limited
Hill & Associates
International Security Academy
Interop
ISI International
International Security Group, ISG
Integrated Convoy Protection
Janusian Security Risk
KBR
Kroll
Lloyd Owens International
L-3 Communications (renamed L3 Technologies)
Majestic Twelve Industries LLC
Meteoric Tactical Solutions
MineTech International
MPRI
Mushriqui Consulting LLC
MVM Inc.
Noble Protective Services
Northbridge Services Group
Nour USA Ltd.
Olive Group
Omega Risk Solutions
Paladin Tactical Solutions
Paratus Worldwide Protection
Parsons
Peak Group Incorporated
Pilgrims Security
Protection Strategies Incorporated
Raymond Associates
RedEye Security and Associates LLC
Reed Incorporated
RGS Logistics
Ronco Consulting Corporation
Safenet Security Services
SAIC, Science Applications International Corporation
Sallyport Global Holding
The Sandi Group
Securiforce International America LLC
Servicio Global de Seguridad e Inteligencia
SkyLink Air and Logistics Support, Inc.
SkyLink Arabia
SOS International, LLC

## List of PMSCs Providing Services in Iraq, 2003–2019*

Spartan Consulting Group
Special Operations Consulting–Security Management Groups, SOC-SMG
SSL (together with Safe Security)
Steele Foundation
Streit Group
Sytex Group, Inc.
TacForce
TigerSwan
Titan Corp
Toifor
TOR International
Total Defense Logistics Ltd.
Trans Atlantic Viking Security, TAV Security
Triple Canopy
United Placements
Unity Resources Group
Universal Security Services LLC
US Investigations Services (USIS)
US Training Center (USTC)
Vinnell
Wamar International LLC
The Whitestone Group
Wolf Group Ltd.
Wolfpack Security & Logistics
Worldwide Language Resources Inc.
Xe
Zapata Inc.
ZKD, LLC

\* companies that meet our definition criteria for a PMSC

## Appendix 4.2  Coding of Variables for Analysis of PMSCs' Human Rights Abuses and Fraud in Iraq

*Dependent Variables:* (1) Fraud. The number of fraud incidents connected to PMSC presence in Iraq. See earlier sections of the Chapter 4 for the definition of fraud and data collection description. (2) Human Rights. The number of human rights violations connected to PMSC presence in Iraq. See earlier sections of Chapter 4 for the definition of human rights and data collection description.

*Key Independent Variable:* Public Ownership. The variable is coded 1 if a company that was present in Iraq at any point was publicly traded and 0 if it was private. If the company's status changed from private to public or vice versa while it was present in Iraq, we noted the transition.

*Control Variables:*

(1) Company Origin. American, European, or other. It is coded according to the origin of the company. These are dummy variables.

(2) Company Size. An ordinal variable coded from 1 to 4, where 1 denotes employment size <1,000, 2 denotes 1,001> and <5,000, 3 denotes >5000 and <10,000, and 4 indicates + 10,000. Larger companies may have more resources to invest in improving employee training on corporate professionalism and on ways to improve interactions with the locals in conflict zones.

(3) Company Age. The variable reports the number of years that have passed since the company's first year of foundation. Companies that have been around for longer may exhibit greater levels of professionalism, with enough experience and resources to improve monitoring mechanisms and establish better corporate discipline to deter crimes.

(4) Subcontracting. We code this 1 if the company subcontracts to other companies, 0 if it doesn't. For example, DynCorp subcontracted to the Sandi Group in 2004.

(5) Membership in Associations. We count the number of memberships in the International Stability Operations Association (ISOA), British Association of Private Security Companies (BAPSC), Security in Complex Environments Group (SCEG), and International Code of Conduct Association (ICoCA).

(6) Services. We categorize services as dummies in the following manner: Intelligence, Training, Demining, Security, Logistics and Base Support, and Other. Other includes maintenance, construction, communication, and technical advising.

(7) Less Competitive & Montreux. In 2008, the Montreux Document established good practices for regulating PMSC operations in armed conflict. Hence, we code each observation as 1 if the company has worked in Iraq during and/or after 2008, the year in which Montreux was signed.

(8) Client Base. We categorize agents in the following manner: Hired by the US government, hired by another government, hired by international organizations, NGOs, or companies. These are coded as dummies.

# 5

# Market Factors and PMSCs' Effectiveness in Conflict Termination

## An Empirical Analysis

The expansion of the PMSC industry has generated interest in gauging the consequences of hiring such companies in war zones where they are involved in a wide range of lethal and non-lethal operations. In this chapter we focus on the impact of PMSCs' interventions on the duration of civil wars from 1990 to 2008. Specifically, we examine whether companies facing a greater level of accountability due to local competition and their publicly traded status are associated with shorter wars in both major and minor civil wars. We argue that multiple security providers hired simultaneously in a given conflict should be more accountable to the client. This accountability, in turn, pushes companies to be more efficient and conscious about adhering to international humanitarian law and anti-fraud practices. The existence of informal rival monitoring mechanism puts pressure on PMSCs to meet the expectations of the client or risk losing the contract to another company. This was the case, for example, when the State Department did not renew the contract of Blackwater, which was implicated in the killings of 17 civilians in 2007, and instead awarded the contract to DynCorp.[1] As such, we expect that as the level of competitive pressure increases, PMSCs will become more accountable and, in turn, contribute to the termination of conflicts.

In the second part of our argument, we posit that some companies have chosen to become publicly traded corporations to secure more capital and, at times, to also credibly signal their commitment to accountability in order to gain an advantage with the rise of global competition among international PMSCs. By becoming publicly traded companies, PMSCs credibly communicate greater commitment to accountability given that such companies are more prone to media and shareholder scrutiny. Whether it is actual misconduct or allegations of misconduct, these can tarnish a company's reputation and survival in a crowded marketplace. Publicly traded companies are thus more vulnerable

*Private Militaries and the Security Industry in Civil Wars.* Seden Akcinaroglu and Elizabeth Radziszewski, Oxford University Press (2020). © Oxford University Press. DOI: 10.1093/oso/9780197520802.001.0001.

to allegations of misconduct and more likely to develop internal monitoring mechanisms to prevent misbehavior by their employees. In 2009, DynCorp, for example, established the position of chief compliance officer, with a specific focus on ethics, business conduct, and regulatory compliance. According to Douglas Ebner, the company's spokesman, DynCorp wanted to signal that the company was "absolutely dedicated to a framework of governance and compliance that ensures a transparent and accountable business environment."[2] Such tactics help create a more skillful and culturally aware workforce and contribute positively to corporate professionalism, factors that improve companies' military effectiveness. Consequently, we expect that interventions by publicly traded PMSCs into civil wars should be associated with greater accountability, more gains for the clients, and, in turn, a quicker termination of conflicts.

## Analysis Overview

As we focus on PMSCs' interventions, our interest is to understand the survival time of civil wars in the world until the event of their termination, whether through a decisive victory, peace agreement, or attrition. Thus, we use a Cox proportional hazards model to estimate the impact of independent variables on the risk of conflict termination in the smallest time span given that the conflict has survived up until that interval.[3]

## Dependent Variable: The Duration of Major and Minor Civil Wars

The duration of civil wars in the world from 1990 to 2008 is our dependent variable. As noted in Chapter 2, we use the Correlates of War (COW) data set (v4.0) and the UCDP/PRIO Armed Conflict data set (v4.0) to include all conflicts in the world that were either still ongoing in 1990 or started during the period 1990–2008. Each observation in the data corresponds to a conflict year, resulting in a total of 321 observations in major conflicts (COW) and 1,327 observations in minor conflicts (UCDP/PRIO). The duration variable is continuous, and conflicts range in duration from a minimum of 1 year to a maximum of 21, years with an average of 5.13 years for COW data. The range is between 1 and 60 years, with an average duration of 12.25 years for the UCDP/PRIO data.

## Key Independent Variables: Local Competition and Corporate Structure

*Competition among PMSCs (Government-Hired PMSCs):* We measure the basic level of local competition among government-hired PMSCs by focusing on the number of PMSCs the government hired in a given conflict. An ordinal variable, it ranges from 0 government-hired PMSCs in a given conflict to 5, the maximum number of providers the government employed in a given conflict for COW data. The maximum number of providers for UCDP/PRIO data (excluding Iraq and Afghanistan) is 21.

*Competition among PMSCs (Rebel-Hired PMSCs):* We measure the basic level of competition among rebel-hired PMSCs by focusing on the number of PMSCs the rebels hired in a given conflict. An ordinal variable, it ranges from 0 or absence of any rebel-hired PMSCs in a given conflict to 5, a maximum number of providers employed by the rebels in a given conflict for COW data. The maximum number of providers in the UCDP/PRIO data is 5.

*Public Ownership:* Using newspaper databases (LexisNexis), scholarly articles, books, blogs, and PMSC websites, we collected information on the corporate structure of PMSCs that have intervened in civil wars from 1990 to 2008. The data indicate whether an intervening company was publicly traded or not.

We also consider alternative factors that could affect the duration of civil wars. These include: *Polity* (the type of governing system in a country), *GDP Per Capita*, *Intensity* (high/low battle deaths in a conflict), *Proportion of Forces* (ratio of government forces to rebel forces), *Mountainous Terrain* (the presence of mountains in a state), *Ethnic Wars* (existence of issues related to ethnicity and ethnic cleavages), *Rebel Support and Government Support* (non-PMSC interventions on behalf of the rebels and/or government), *Ethnic Fractionalization* (level of ethnic division), and *Population* (size of the state's population).[4]

## Findings: Local Competition, PMSCs' Interventions, and the Duration of Civil Wars

We find support for our argument that local competition improves PMSCs' performance in and brings positive developments toward conflict termination in major wars (COW data set) (Table 5.1). By shifting

Table 5.1  PMSC's interventions, local competition, and the duration of civil wars in the world, 1990–2008, major conflicts

|  | (1) Model 1 | (2) Model 2 | (3) Model 3 |
|---|---|---|---|
| GDP Per Capita | 0.92* | 0.92* | 0.92* |
|  | (0.05) | (0.04) | (0.05) |
| EthnicWar | 1.54 | 1.52 | 1.61 |
|  | (0.54) | (0.50) | (0.56) |
| Intensity | 0.82*** | 0.82*** | 0.81*** |
|  | (0.03) | (0.03) | (0.03) |
| Mountainous Terrain | 0.88*** | 0.88*** | 0.88*** |
|  | (0.03) | (0.03) | (0.03) |
| Polity | 1.00 | 1.00 | 1.00 |
|  | (0.01) | (0.01) | (0.01) |
| Proportion of Forces | 0.96 | 0.97 | 0.97 |
|  | (0.06) | (0.06) | (0.06) |
| Population | 0.91 | 0.86 | 0.90 |
|  | (0.12) | (0.11) | (0.12) |
| Support Rebels | 0.44* | 0.48 | 0.44* |
|  | (0.21) | (0.23) | (0.21) |
| Support Government | 1.03 | 1.00 | 1.11 |
|  | (0.47) | (0.43) | (0.50) |
| Ethnic Fractionalization | 0.73* | 0.77 | 0.73* |
|  | (0.12) | (0.13) | (0.12) |
| Competition PMSC Gov | 1.41*** |  | 1.37*** |
|  | (0.16) |  | (0.15) |
| Competition PMSC Reb |  | 1.59** | 1.42** |
|  |  | (0.33) | (0.23) |
| *Time Varying Covariates* |  |  |  |
| Support Rebels | 1.27*** | 1.26** | 1.27*** |
|  | (0.11) | (0.12) | (0.11) |
| Polity | 1.00 | 1.00 | 1.00 |
|  | (0.00) | (0.00) | (0.00) |
| N | 319 | 319 | 319 |

Exponentiated coefficients; Standard errors in parentheses;
*p<0.1, **p<0.05, ***p<0.01

the balance of power in favor of the client, PMSC interventions may help one party achieve an outright victory or put pressure on the enemy to negotiate or risk an impending defeat. We report the hazard rates of each variable in Table 5.1, where the hazard rate denotes the risk of termination of conflict. Any variable with the hazard rate exceeding 1 contributes to the termination of conflict, while variables with hazard rates that are smaller than 1 prolong the duration of conflicts. Thus, Models 1–3 show robust support for the relationship between PMSC competition and termination of wars when governments outsource security needs to such actors. The hazard rates for our main variable of interest, *Competition PMSC Gov*, are 1.41 in Model 1 and 1.37 in Model 3. Because it exceeds 1, the hazard rate shows that as the number of government-hired PMSCs increases in a given conflict, the termination of such conflicts is more likely. There is thus a substantial difference in the duration of conflicts with and without local competition for government-hired PMSCs.

An increase in competition among rebel-hired PMSCs is also linked to shorter wars (variable *Competition PMSC Reb*, Models 2 and 3). Greater accountability among a larger number of companies supporting the rebels is likely to increase the rebels' power and push the government to negotiation rather than to seek out an all-out victory and prolong the war. However, the results related to PMSCs hired by the rebels need to be interpreted with caution as we have only seven conflicts in our data with PMSC intervention on behalf of these non-state actors. Given the stigma associated with working for the rebels, we may lack additional data in the future to derive robust findings about the impact of competition on the effectiveness of rebel-hired international PMSCs. As such, our analysis here focuses more in depth on the relationship between market forces, government-hired international PMSCs, and conflict duration. Future analysis might consider the impact of mercenaries and quasi-companies that fight on behalf of the rebels.

Figure 5.1 delves deeper into the relationship between local competition among government-hired PMSCs and the duration of war. On the $x$-axis, we plot the duration of conflicts in years, while on the $y$-axis we plot the survival rate, which captures the probability that war continues until that time period. In the absence of competition for PMSCs hired by the government (CompPMSCGov=1), the survival rate for a conflict that has been ongoing for five years is approximately 0.8 and not much different from the survival rate in the absence of any PMSC intervention (CompPMSCGov=0). However, once the number of security providers increases, the survival rate for the conflict decreases substantially (around 0.5). In other words, the competitive environment is having a significantly larger effect on decreasing

Cox proportional hazards regression

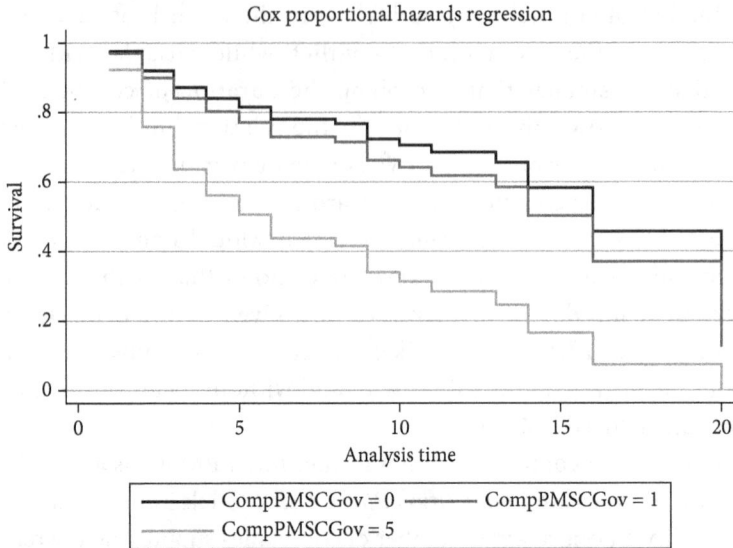

**Figure 5.1** Survival function for the termination of conflict in the presence and absence of local competition for government-hired PMSCs in major conflicts
*Note*: Generated from Table 5.1, Model 3 without the time varying covariates.

the conflicts' survival rates relative to cases when a PMSC is a sole security provider.

One may argue that our findings do not necessarily support the argument that competitive pressure pushes PMSCs to be more accountable and thus more likely to contribute to shorter wars, but that our results are driven by a simple accumulation of gains that would be expected from multiple providers. In short, the more PMSCs the government hires, the more services it gets. To assess the independent and additional effect of PMSCs' accountability, we conducted an analysis of the duration of war in which we controlled for the total number of services provided by existing PMSCs. It should not be assumed that the number of services is greater with more PMSC interventions. Our data show that it is not uncommon for a single PMSC to deliver multiple services yet operate as a sole provider without competitive oversight that comes with multiple interventions. Table 5.2 shows that *Services*, an ordinal variable that captures the total number of services delivered by all PMSCs that have intervened in a given conflict year, is not statistically significant in shaping the duration of war in major conflicts, but the competitive pressure is. The competitive pressure increases the accountability of PMSCs and in doing so pushes them to deliver higher-quality services and operate more in accordance with international humanitarian law.

**Table 5.2** Number of PMSC services, competitive pressure, and the duration of civil wars in the world, 1990–2008, major conflicts

|  | (1)<br>Model 1 |
| --- | --- |
| GDP Per Capita | 0.92*<br>(0.04) |
| Ethnic Wars | 1.51<br>(0.53) |
| Intensity | 0.81***<br>(0.03) |
| Mountainous Terrain | 0.89***<br>(0.03) |
| Polity | 1.00<br>(0.01) |
| Proportion of Forces | 0.97<br>(0.06) |
| Population | 0.89<br>(0.12) |
| Support Rebels | 0.45*<br>(0.22) |
| Support Government | 1.01<br>(0.45) |
| Ethnic Fractionalization | 0.74*<br>(0.12) |
| Competition PMSC Gov | 3.45*<br>(2.44) |
| Services | 1.02<br>(0.09) |
| *Time Varying Covariates* |  |
| Support Rebels | 1.27***<br>(0.11) |
| Polity | 1.00<br>(0.00) |
| N | 319 |

Exponentiated coefficients; Standard errors in parentheses
* p<0.1, ** p<0.05, *** p<0.01

Therefore, the results in Table 5.2 illustrate that the causal story of multiple PMSCs' interventions and their impact on war termination is more about accountability spurred by multiple interventions than merely about accumulation of services (resources). The war is likely to end sooner when multiple PMSCs intervene not because the governments are getting more assistance, but because the contractors are delivering a better quality of services. By refraining from harassing the locals, avoiding mismanagement of funds, and engaging the enemy as specified by the contract, they are able to push the balance of power in favor of the government to help secure an outright victory or put pressure on the rebels to negotiate. When it comes to the success of PMSCs' interventions, quality does trump quantity.

So far we have examined whether an overall increase in PMSCs' interventions, the rise in local competition, creates more accountability among providers and in doing so makes them more likely to contribute to the termination of conflicts. We can also analyze whether this competitive pressure operates in the same way regardless of the type of services that companies deliver. To do so, we coded data on the types of services that PMSCs have provided to governments based on four categories: Direct/indirect combat involvement (variable *Combat* in Table 5.3), intelligence gathering and communications (*Communications* variable in Table 5.3), consultancy and training of police and troops (*Training* variable in Table 5.3), and asset and site security (*Security* variable in Table 5.3). For each conflict year in the context of major wars, we coded the number of companies that the government has hired to provide support in the same service category, for example, all companies that have assisted in combat operations in a given conflict year.

Our analysis shows that with the exception of intelligence gathering and communications, which is not statistically significant, an increase in competition in each service category puts pressure on companies to step up and match their behavior with the expectations of the client. Once this alignment occurs, the companies are in a better position to be more effective in contributing to conflict termination. What this analysis shows is that whether we adapt a more aggregate measure of competition by looking at all PMSC interventions in a given year regardless of the type of service they provide or consider competition only in a specific service category, the results are almost the same. Both measures support our earlier finding that it is the quality of the service that matters more than type (with the exception of intelligence gathering and communications) and quantity. The statistical significance of the "competition" variable shows that having more providers increases PMSCs' military effectiveness.

Does the competitive pressure faced by PMSCs exert the same positive impact on the termination of minor conflicts? Using the UCDP/PRIO Armed Conflict data set, we note 1,326 observations, with 116 episodes of conflict termination. In Models 1–3 in Table 5.4, we test the claim that

**Table 5.3** PMSCs' interventions, competition in different service categories, and the duration of civil wars in the world, 1990–2008, major conflicts

| | (1) Model 1 | (2) Model 2 | (3) Model 3 | (4) Model 4 |
|---|---|---|---|---|
| GDP Per Capita | 0.92* | 0.92* | 0.92* | 0.92* |
| | (0.05) | (0.05) | (0.04) | (0.04) |
| Ethnic Wars | 1.66 | 1.49 | 1.49 | 1.52 |
| | (0.57) | (0.52) | (0.51) | (0.53) |
| Intensity | 0.81*** | 0.81*** | 0.82*** | 0.82*** |
| | (0.03) | (0.03) | (0.03) | (0.03) |
| Mountainous Terrain | 0.88*** | 0.88*** | 0.88*** | 0.88*** |
| | (0.03) | (0.03) | (0.03) | (0.03) |
| Polity | 1.00 | 1.00 | 1.00 | 1.00 |
| | (0.01) | (0.01) | (0.01) | (0.01) |
| Proportion of Forces | 0.97 | 0.96 | 0.96 | 0.96 |
| | (0.06) | (0.06) | (0.06) | (0.06) |
| Population | 0.89 | 0.89 | 0.88 | 0.89 |
| | (0.12) | (0.12) | (0.11) | (0.12) |
| Support Rebels | 0.42* | 0.49 | 0.47 | 0.47 |
| | (0.20) | (0.24) | (0.23) | (0.23) |
| Support Government | 1.14 | 1.01 | 0.96 | 0.97 |
| | (0.51) | (0.46) | (0.43) | (0.43) |
| Ethnic Fractionalization | 0.75* | 0.74* | 0.77 | 0.75* |
| | (0.12) | (0.12) | (0.13) | (0.12) |
| Training | 1.63*** | | | |
| | (0.17) | | | |
| Combat | | 1.37*** | | |
| | | (0.14) | | |
| Communications | | | 1.46 | |
| | | | (0.46) | |
| Security | | | | 1.51** |
| | | | | (0.29) |
| *Time Varying Covariates* | | | | |
| Support Rebels | 1.27*** | 1.26** | 1.27*** | 1.27*** |
| | (0.11) | (0.11) | (0.12) | (0.11) |
| Polity | 1.00 | 1.00 | 1.00 | 1.00 |
| | (0.00) | (0.00) | (0.00) | (0.00) |
| N | 319 | 319 | 319 | 319 |

Exponentiated coefficients; Standard errors in parentheses; * p<0.1, ** p<0.05, *** p<0.01

Table 5.4 PMSCs' interventions, local competition, and the duration of civil wars in the world, 1990–2008, minor conflicts

| | (1) Model 1 | (2) Model 2 | (3) Model 3 |
|---|---|---|---|
| GDP Per Capita | 0.96 (0.03) | 0.96 (0.03) | 0.96 (0.03) |
| Ethnic Wars | 1.27 (0.28) | 1.19 (0.27) | 1.26 (0.28) |
| Intensity | 0.19*** 0.06 | 0.20*** 0.07 | 0.19*** 0.06 |
| Mountainous Terrain | 1.00 (0.04) | 0.99 (0.04) | 1.00 (0.04) |
| Polity | 1.00 (0.01) | 1.01 (0.01) | 1.00 (0.01) |
| Proportion of Forces | 0.87*** (0.03) | 0.87*** (0.03) | 0.87*** (0.03) |
| Population | 0.78*** (0.06) | 0.79*** (0.06) | 0.78*** (0.06) |
| Support Rebels | 0.39*** (0.10) | 0.40*** (0.11) | 0.39*** (0.10) |
| Support Government | 0.33*** (0.06) | 0.35*** (0.07) | 0.33*** (0.06) |
| Ethnic Fractionalization | 1.39*** (0.16) | 1.35*** (0.16) | 1.39*** (0.17) |
| Competition PMSC Gov | 0.75** (0.10) | | 0.75** (0.10) |
| Competition PMSC Reb | | 0.81 (0.25) | 0.89 (0.22) |
| *Time Varying Covariates* | | | |
| Polity | 1.00*** (0.00) | 1.00*** (0.00) | 1.00*** (0.00) |
| Support Rebels | 1.01 (0.03) | 1.02 (0.04) | 1.01 (0.03) |
| MountainTerrain | 0.98*** (0.01) | 0.98*** (0.01) | 0.98*** (0.01) |
| N | 1,326 | 1,326 | 1,326 |

Exponentiated coefficients; Standard errors in parentheses * $p<0.1$, ** $p<0.05$, *** $p<0.01$; one missing observation

local competition in security market matters in the termination of minor conflicts. While we find statistical significance for local competition when governments hire PMSCs (Models 1 and 3), the direction of the relationship runs contrary to our expectations. Local competition is associated with prolonged minor conflicts; in fact, conflict termination is 25% less likely (Models 1 and 3) when multiple PMSCs operate in minor conflict zones. Competition among PMSCs' hired by the rebels is not statistically significant and also runs contrary to the expected direction. This shows a different dynamic of PMSCs' interventions in major vs. minor conflicts.

This finding, however, does not immediately imply that competitive pressure has the opposite effect in minor conflicts. The different complexity of major and minor conflicts and the selection bias likely explain why competitive pressure faced by PMSCs is linked to shorter conflicts in major wars and to longer ones in minor wars. Paradoxically, it is often the case that minor conflicts are harder to resolve compared to major conflicts. In most instances of minor conflicts, rebel groups adopt insurgency and terrorist tactics rather than conventional battle strategies,[5] while in major conflicts strategies are mixed and may involve stronger rebel movements with the capacity to confront the government more openly. By contrast, insurgents in minor conflicts tend to run or hide among the civilians and complicate the effort to neutralize the threat. Even though PMSC services have diversified to accommodate the needs of their clients in minor conflict zones, it is a lot harder to bring conflicts to an end when PMSCs and government forces are unable to confront rebels in a conventional way. Indeed, average conflict duration in the minor conflicts data set is approximately 12 years, as opposed to 5 years in major conflicts. While this is partially due to the way thresholds are defined for civil wars—wars are coded as terminated when the 1,000 annual battle death threshold is not met in major wars versus 25 battle deaths in minor wars—it can also be attributed to the different strategies employed by non-state challengers in the two contexts. Consequently, this suggests that even when competitive pressure pushes PMSCs to be more accountable, the increase in accountability in service provision is not sufficient to bring an end to hostilities in minor conflicts. The findings for minor conflicts therefore do not imply that competitive pressure is not important in conflict dynamics; rather, they show that even with this mechanism in place, the strength of PMSC interventions is not sufficient to terminate this type of conflicts, at least in the short term.

It is not just that minor conflicts in general are more difficult to end compared to major conflicts, but it is also the case that PMSCs are more likely to be hired in minor conflicts that are especially hard to resolve, suggesting the presence of a selection bias. For example, in Colombia and Iraq, several insurgent groups have targeted the government; in those cases insurgents use terror tactics and rely on resources such as oil and precious minerals that can sustain their fighting. It is access to such resources that is associated with prolonged wars in the literature.[6] The absence of conflict termination in these and similar conflicts should not be automatically construed as the outcome of suboptimal performance by PMSCs or as evidence that local competition as a monitoring tool is ineffective in these conflict zones. If our argument were correct, then the counterfactual, the absence of PMSC competition in minor conflicts, would have driven these conflicts to be even longer.

Overall, it is the most likely case that these wars are long not because of competitive PMSC intervention but in spite of it. Thyne,[7] for example, demonstrates that a similar selection bias explains the relationship between UN interventions and longer wars. He shows that UN interventions are associated with long wars, though not because these interventions somehow alter the war's dynamics in a negative way but rather because the UN intervenes in those conflicts that are notoriously difficult to resolve and that have been ongoing for years. Our findings about PMSCs' impact on the duration of minor conflicts follow the same logic. If a selection bias exists in the case of PMSCs' interventions in minor conflicts, it means that multiple PMSCs are hired in extremely difficult conflict zones and thus we see that competition, the presence of multiple PMSCs, is associated with prolonged minor conflicts.

To examine this possibility, we run a statistical analysis to examine whether multiple PMSCs are more likely to intervene on behalf of the governments in hard-to-resolve civil wars. Civil war literature shows that the presence of multiple rebel groups[8] that are relatively strong[9] with access to natural resources,[10] as well as insurgencies where terrorist tactics are the defining feature of the conflict,[11] make conflicts particularly hard to end. We adopt these as independent variables and run a logistic model.[12] The results show that in the cases of minor conflicts, clients are more likely to hire multiple PMSCs in conflict zones where several insurgent groups are fighting (Table 5.5). In conflicts with several insurgent groups, commitment and informational problems tend to be acute[13] in ways that prevent accurate estimation of capabilities so vital to successful bargaining and conflict termination. The expectation that at least one group will renege on the agreement also creates commitment issues. Thus, wars with multiple groups are notoriously difficult to resolve.

Second, the results also indicate that multiple PMSCs are more likely to intervene in conflicts where groups use terrorist tactics. Because it is extremely

Table 5.5 Multiple PMSCs and hard-to-resolve
conflicts: selection bias test

|  | (1)<br>Model 1 |
| --- | --- |
| Proportion of Forces | −0.07*<br>(0.04) |
| Resources | 0.94***<br>(0.13) |
| Terrorist Attacks | 0.003***<br>(0.00) |
| Number of Groups | 0.41***<br>(0.12) |
| N | 1,324 |

Constant is not reported, Standard errors in parentheses
* p<0.1, ** p<0.05, *** p<0.01; three missing observations

difficult to pinpoint the exact location of terrorists, such groups are harder to defeat. By using surprise attacks, terrorists can also inflict heavy damages either indirectly through civilian casualties or directly through their attacks on military facilities.

Third, clients rely on multiple PMSCs when they face a particularly strong challenger; that is, in cases where the proportion of government to rebel forces is lower. Lastly, the probability of hiring multiple PMSCs in minor conflicts is higher in the presence of resources. Resources delay conflict termination in two ways, by providing rebels with a source of funding to continue their struggle and/or because they create the so-called resource curse, a phenomenon where rebels are unwilling to negotiate and lose their profit-making enterprise from controlling resource-rich areas.[14] Combined, these results demonstrate that multiple PMSCs—the presence of a competitive PMSC environment—are likely to be hired in protracted conflicts, and we should be aware of this selection bias when evaluating PMSCs' performance in minor conflicts. Because PMSCs face particularly acute challenges when they are hired in minor conflicts, they are unable to bring the conflict to a swift end. This, however, does not mean that local competition has a negative impact and that it should not be encouraged. It is possible that without the presence of multiple PMSCs, the termination of these hard-to-resolve conflicts would have been even more problematic.

## Findings: Corporate Structure, PMSCs' Interventions, and the Duration of Civil Wars

Testing our argument regarding the impact of PMSCs' corporate structure in the context of major conflicts was not an option, as there are only two cases of intervention involving publicly traded companies, Defense Systems Limited, hired in the Democratic Republic of the Congo in 1998 and DynCorp hired in Pakistan in 2006. In minor conflicts, we have a larger range of conflicts (13) where publicly traded PMSCs have provided services. The focus is thus on minor conflicts, where we find support for our expectation (Table 5.6). The hazard rates for the key independent variable, *Public Ownership*, exceed 1 in both models, showing that publicly traded PMSCs are associated with shorter conflicts. Specifically, the results show that the presence of publicly traded PMSCs is linked with a greater than twofold (Model 1) and sevenfold (Model 2) higher incidence of conflict termination in comparison to cases when only private companies are hired.[15]

We delve deeper into the impact of interventions by publicly traded PMSCs by plotting the duration of conflicts in years on $x$-axis and the survival rate on the $y$-axis. Figure 5.2 shows, for example, that the survival rate for a conflict in its 10th year in the absence of interventions by publicly trade PMSCs (Publiclyowned=0) is high at approximately 0.96, while the survival rate drops to around 0.78 in those cases when publicly trade PMSCs intervene (Publiclyowned=1). This shows the positive impact of publicly traded PMSCs' on military effectiveness.

How do we account for the finding that interventions involving publicly traded PMSCs in minor conflicts are linked to shorter wars but interventions by multiple PMSCs are linked to longer wars in the context of minor conflicts? We noted that governments usually work with multiple PMSCs in the most challenging minor conflicts; PMSCs do not prolong the wars but rather they are linked to longer wars because of a selection bias. Difficult conflicts push the governments to seek help from multiple companies. Could a selection bias also exist in those cases where publicly traded PMSCs intervene? In other words, is it the case that publicly traded companies are linked to shorter civil wars in the context of minor conflicts because they are strategic in their selection of clients? As publicly traded companies are more vulnerable to reputational costs for limited effectiveness, they may be choosing to bid for contracts where they estimate the highest level of success. Findings from Table 5.7 lend some support to this idea. A logit model analyzes factors that affect interventions by publicly traded PMSCs. Two factors associated with hard-to-end conflicts, an increase in the number of rebel groups and

**Table 5.6** PMSCs' corporate structure and the duration of civil wars in the world, 1990–2008, minor conflicts

|  | (1) Model 1 | (2) Model 2 |
|---|---|---|
| GDP Per Capita | 0.96 (0.03) | 0.96 (0.03) |
| Ethnic Wars | 1.30 (0.29) | 1.36 (0.30) |
| Intensity | 0.21*** 0.07 | 0.21*** 0.07 |
| Mountainous Terrain | 0.93** (0.03) | 0.92** (0.03) |
| Polity | 1.00 (0.01) | 1.00 (0.01) |
| Proportion of Forces | 0.88*** (0.03) | 0.88*** (0.03) |
| Population | 0.78*** (0.06) | 0.78*** (0.06) |
| Support Rebels | 0.40*** (0.11) | 0.38*** (0.10) |
| Support Government | 0.35*** (0.07) | 0.33*** (0.06) |
| Ethnic Fractionalization | 1.27** (0.14) | 1.27** (0.14) |
| Public Ownership | 2.65** (1.13) | 7.84*** (4.85) |
| Competition PMSC Gov |  | 0.60* (0.18) |
| Competition PMSC Reb |  | 0.95 (0.22) |
| *Time Varying Covariates* |  |  |
| Polity | 1.00*** (0.00) | 1.00*** (0.00) |
| Support Rebels | 1.03 (0.04) | 1.02 (0.03) |
| N | 1,326 | 1,326 |

Exponentiated coefficients; Standard errors in parentheses

* p<0.1, ** p<0.05, *** p<0.01; one missing observation

an increase in terror attacks, are linked to interventions by publicly traded PMSCs. However, the presence of natural resources and proportion of government to rebel forces, the phenomena that matter in prolonging civil wars, are not statistically significant and do not account for interventions by

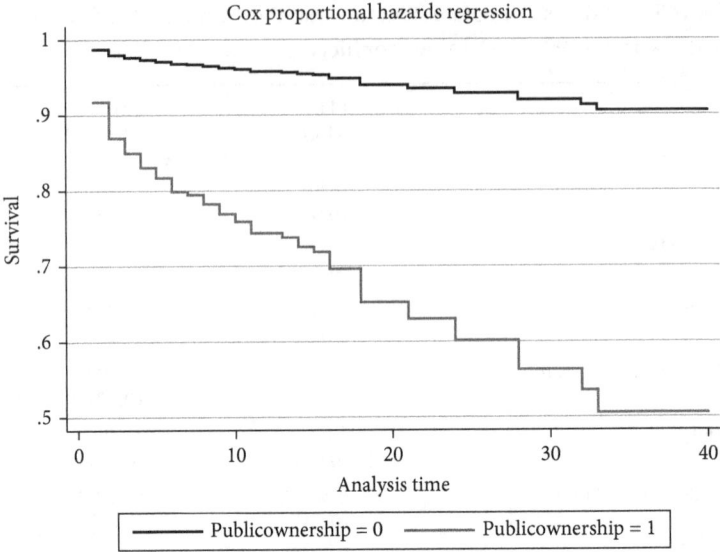

**Figure 5.2** PMSCs' corporate structure and conflict termination, minor conflicts
*Note*: Generated from Table 5.6, Model 2 without the time varying covariates.

**Table 5.7** Publicly traded companies and hard-to-resolve conflicts: selection bias test

|  | (1)<br>Model 1 |
| --- | --- |
| Proportion of Forces | 0.46 |
|  | (0.05) |
| Resources | 0.35 |
|  | (0.22) |
| Terrorist Attacks | 0.01*** |
|  | (0.00) |
| No of Groups | 0.53*** |
|  | (0.19) |
| N | 1,324 |

Constant is not reported; Standard errors in parentheses * p<0.1, ** p<0.05, *** p<0.01; three missing observations

publicly traded PMSCs. This might suggest that publicly traded PMSCs may be more averse to interventions in the most complex conflict zones but not necessarily in the somewhat difficult ones.

A clearer picture emerges when we calculate predicted probabilities for interventions by publicly traded PMSCs into hard-to-resolve conflicts and compare those with predicted probabilities for interventions into the same type of conflicts by private PMSCs. The probability that a publicly traded PMSC will intervene in a very difficult minor conflict zone, for example, one with five rebel groups, 200 annual terror attacks, and one that has abundant natural resources, is 0.39 while an intervention by a private PMSC is much higher at 0.54. What this shows is that while publicly traded companies do intervene in hard-to-resolve conflicts, they are more likely than a private PMSC to avoid working in the most challenging conflicts. Publicly traded companies are vulnerable to shareholder reactions to negative publicity and, given that failure to deliver successful results increases in the most complex environments, such companies might strategically avoid working for clients that face a low possibility of military success to begin with.

## Additional Factors Affecting the Termination of Civil Wars

So far we have examined the impact of local competition and corporate structure on PMSCs' performance. In our analysis we also considered other factors commonly cited in the civil war literature as important in shaping the duration of wars. Our findings differ for major and minor conflicts. Conflicts are likely to be prolonged when each of the combatants receives external aid in minor and major conflicts.[16] Furthermore, in conflicts, the impact of time has a statistically significant impact on how support to rebels shapes the war dynamics, and this finding is robust across all models. With the passage of time, external support to such non-state actors increases the odds of war termination because the warring parties can now observe that no one can achieve an outright victory and this pushes the groups to negotiate.

We have a counterintuitive finding about the proportion of forces of combatants, the ratio of the military forces over the rebel forces. This variable was statistically insignificant in the major conflicts but matters in minor civil wars. This finding holds for all models; however, the magnitude of the coefficient is unexpected. The results show that high government capabilities relative to rebel forces tend to delay the termination of conflict. It is likely the case that in minor conflicts weaker rebels have the means to survive, either because they do not engage in face-to-face confrontations with the government and rely heavily on guerrilla tactics or because capable governments may be underestimating the threat coming from weaker rebels and hence lack the urgency to immediately defeat such groups.[17]

We note other findings unique to minor conflicts. In such contexts democracies are facing slightly longer wars as time goes by (statistical significance of the time-varying covariate *Polity* with the coefficient rounded up to 1). This is a surprising result as we expected democracies to have greater state capacity and strong norms of accommodation that would result in quicker termination of wars. However, given that democracies face audience costs for giving too many concessions to groups, democratic governments may be averse to negotiating with the rebels if the conflict continues and instead push for a victory. Minor conflicts are also more likely to be linked to longer wars when population size increases. Larger populations provide more opportunities for the rebels to recruit supporters and mount a formidable challenge against the government. Finally, we note that ethnic fractionalization reduces the duration of war because more ethnically divided societies make it harder for the rebels to recruit supporters. Linguistic differences as well as more diverse set of grievances among different groups might also limit the rebels' ability to maintain group cohesion. These developments are likely to strengthen the government's power over the rebels.

A finding that is specific to major wars shows that higher levels of GDP per capita are connected to slightly longer wars across all models. This result is unexpected as higher GDP per capita should be linked to greater state capacity to effectively manage the war. While higher GDP reduces a state's risk of experiencing a civil war, it is also possible that when faced with rebel challenge, more prosperous governments will overestimate their military ability and aim for an all-out victory. As they resist early attempts to negotiate the result could be a more prolonged war.

We note some similarities in both major and minor conflicts when it comes to conflict duration. Specifically, having large mountainous terrain provides favorable conditions for the rebels to survive, and is associated with longer wars, though in minor wars this effect becomes stronger over time. More-intense conflicts or those with large battle deaths are also linked to prolonged fighting in both contexts. Though intense wars can push the warring parties to a mutually hurting stalemate that is conducive to negotiations, they may also lead to the accumulation of grievances that increases rebel resolve.

## Comparison of PMSCs and Non-PMSC Interventions

Given the existence of extensive literature on outside interventions into civil wars,[18] it is worth examining how interventions by PMSCs compare

to other types of external interventions in their impact on conflict duration. Our findings indicate that under specific conditions, interventions
by PMSCs on behalf of the government can bring positive contribution
to the termination of conflicts. PMSCs do not intervene nearly as frequently into civil wars as states and international organizations such as
the UN. For example, their assistance is noted in only 9% of minor conflict
years, in comparison to interventions by states and international organizations that are present in 87% of the conflict years. As such, concerns
about international PMSCs taking over control of states' monopoly on the
use of force are not supported by the data. This does not exclude the possibility that conflict zones create lucrative opportunities for mercenaries
and quasi-companies to challenge state power especially in the context
of weak states. Our data show that simultaneous interventions are rare.
PMSCs have assisted governments in the presence of other intervening
actors in 7% of the conflict years.

Figure 5.3 illustrates what happens to the termination of minor conflicts
depending on who the intervening actors are. It compares the survival rate
in two contexts: when local competition exists and two PMSCs are the
only intervening agents, and when the government hires only one PMSC

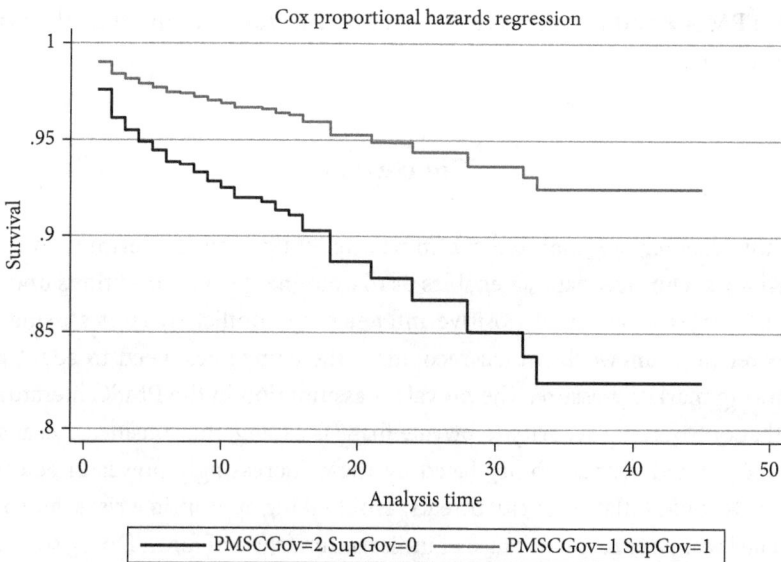

Figure 5.3 The impact of external interventions and multiple PMSCs'
interventions (presence of local competition) on the duration of minor conflicts

(absence of local competition) in addition to receiving external non-PMSC support. The survival rate for conflict is slightly lower during any year of conflict when multiple PMSCs intervene on behalf of the government in the absence of non-PMSC intervention, as opposed to when non-PMSC interventions occur along with only one PMSC. While multiple PMSCs are more likely to intervene in hardest-to-resolve wars where they are not as effective in ending the wars due to conflict complexity, nevertheless governments might be better off relying on multiple PMSCs rather than on external interventions by states/international organizations if they want to shorten the durations of wars. External sponsors' agenda may not completely overlap with the interests of the government, and when intervention is limited in scope and duration, it may not be adequate to bring major gains. States and international organizations face greater limitations on flexibility, while PMSCs can enter the war zone much faster. Most notable here is an insight that adding any additional actors into the conflict zone does not automatically generate the informal monitoring mechanism that creates accountability for the PMSC industry. Rather, there is a strong likelihood that the mechanism operates when additional actors are PMSCs themselves. This finding is consistent with the logic of market pressure exerted by players from the same industry. Here again, states and international organizations may have very different agendas from PMSCs and not necessarily encourage its forces to monitor the private actors.

## Conclusion

In this chapter, we analyzed the impact of PMSCs on the termination of civil wars. Our new data set enables us to examine specific conditions under which PMSCs can exert positive influence on conflict dynamics using a theoretical framework, which recognizes the companies' need to adapt to growing market pressure. The prevalent assumption in the PMSC literature is that outsourcing security to private firms is a dangerous business because of the limited accountability faced by these increasingly powerful actors. PMSCs could inflate security threats, avoid taking appropriate risks, and alienate local population because of indiscriminate use of force. Doing so may allow the companies to secure more profits as long as such behavior remains undetected. Because of such concerns, the dominant view among scholars and NGOs is that PMSCs may be unreliable agents of security. Our analysis shows that PMSCs can pursue more aggressive self-monitoring behavior under the right conditions. When this occurs, PMSCs can have a more positive impact on the termination of civil wars than interventions by states and international organizations. In fact, when given a choice, governments should turn to multiple PMSCs for assistance rather than seek external assistance from other actors.

How can governments harness the power of PMSCs in a way that benefits them the most? Our findings show that when the governments rely on more than one company for services, PMSCs' interventions are associated with shorter civil wars in the context of major conflicts. The presence of competition in a given conflict puts pressure on companies to embrace a more accountable strategy in accordance with the clients' expectations. For example, companies might avoid mismanagement of funds and use the resources to help shift the balance of power in favor of the government. Increasingly, international efforts by NGOs and scholars have raised awareness among many governments about the need to specifically mention adherence to international humanitarian law in contracts. This recognition is not only beneficial for ethical reasons. Human rights abuses committed by government forces and private contracts often prompt the locals to side with the rebels, inhibiting the government's ability to end the war. Our findings show that governments hoping to benefit the most from PMSCs' assistance should divide contracts among several providers since competitive pressure resulting from the presence of multiple providers increases monitoring among those providers.

Second, our analysis shows that the industry's expansion heralds some positive developments for global security. With growing competition, companies have restructured their corporate approach, some becoming publicly traded not only to raise capital but also to signal to clients their commitment to transparency and greater accountability. These companies have not been present in any major conflicts, but they have played a role in minor conflicts with promising results. Our findings show that in contexts where publicly traded PMSCs intervene to assist the governments, they shift the balance of power in favor of the client, thereby increasing the probability of war termination. These companies are also somewhat more likely to intervene in those minor conflicts that are not the most difficult to resolve, possibly selecting strategically to intervene in places where they estimate their likelihood of success to be large. Because publicly traded companies are much more accountable to shareholders, they have more to lose from failing to deliver to the client.

Hard-to-resolve conflicts, such as those where several rebel groups challenge the government, where such rebels have access to resources including oil and precious minerals to sustain their fight or employ terrorist tactics, are a challenge even for multiple PMSCs. Findings in this chapter show that this type of conflict is more likely to involve multiple PMSC interventions. These interventions, while not bringing an immediate end to hostilities, are still more promising than interventions by other actors in improving the gains for the government. Put simply, when faced with hard-to-resolve conflicts, governments must be realistic in their expectations and note that with competitive pressure increasing PMSCs' accountability, the most difficult conditions on the ground will be challenging even for those private contractors that are committed to fulfilling the terms of the contract. Naturally, this prompts us to ask whether the governments should even turn to PMSCs in hard-to-resolve conflicts if the termination of war appears distant. While such interventions do not deliver immediate gains, they may help the government survive long enough to diminish rebel resolve. In the absence of such interventions, rebel resolve may increase and the possibility of negotiations could become even more remote.

The PMSC industry is continuously expanding and the need for PMSCs is unlikely to abate soon. In this chapter, we showed that under specific conditions, the alignment between profit-seeking companies and the client is possible. The distribution of security contracts among multiple providers creates an incentive for PMSCs to deliver better outcomes, while public ownership puts pressure on companies to appease shareholders by building a positive reputation for military effectiveness.

## Appendix 5.1  Control Variables, Selection Bias, and Robustness Tests for Empirical Analysis of PMSCs' Interventions into Civil Wars

### Control Variables for the Analysis

*Proportion of Forces*: The variable represents government forces divided by rebel forces. The data for military personnel until 2001 come from the COW Material Capabilities data set. The data between 2001 and 2008 comes from World Bank Development Indicators. The correlation between the two datasets is 0.91 and thus any potential bias from merging them is likely to be random. We obtained the data on rebel forces from Cunningham et al. (2009)[19] and updated it for missing values and for recent wars. We expect the duration of war to be shorter when the proportion of government forces to those of the rebels is larger. This is in line with Fearon (2004),[20] who argues that wars are shorter when conditions favor a decisive victory. Similarly, Herbst (2004)[21] points out that many African civil wars are long because of the inadequate military mobilization of African armies. This suggests that strengthening government capabilities may shorten wars. We took the log transformation of the variable to avoid skewness.

*GDP Per Capita:* This variable captures real Gross Domestic Product per capita adjusted for prices in 2005 and comes from Penn World Table v. 6.3.[22] GDP per capita is a robust finding in the civil war literature; not only is GDP per capita associated with higher state capacity[23] but also higher opportunity costs for rebellion, which makes rebel recruitment an expensive deed.[24] Thus, higher GDP per capita should act as a deterrent against longer wars. We took the log transformation of the variable to avoid skewness.

*Polity:* Polity captures the type of governing system in a country and ranges from −10, a mature authoritarian regime to 10, a mature democratic regime. We obtain the scores from Polity IV Project, 1800–2013.[25] Opportunity costs for a rebellion are likely to be higher in democratic societies, as rebels will have alternative means of resolving contentious issues. Norms of accommodation and peaceful conflict resolution will be valued in such places, prompting leaders to take initiatives to end wars. Furthermore, Balch-Lindsay and Enterline (2000)[26] argue that in mature democracies, local bureaucracies are more efficient and experienced in dealing with difficult policy questions, and better equipped to resolve contentious issues that might be arising between parties. This leads us to believe that mature democracies are more capable of bringing a resolution to conflicts.

*Ethnic Wars:* The variable denotes the type of issues involved in civil wars and comes from Cunningham et al. (2009).[27] Many scholars argue that ethnic conflicts may display an "all-or-nothing" character and may have attributes that differ from other types of war.[28] Ethnic issues are harder to resolve because nationalist rhetoric hardens group cleavages, making it difficult for intergroup dialogue to emerge. This leads to greater distrust and intensifies the security dilemma, all of which may prolong the conflict. Kaufmann (1996),[29] for example, argues that partition might be the only solution to resolving ethnic wars. We code ethnic wars as a dummy.

*Intensity:* Coded either as more intense or less intense and based on the number of battle deaths in armed conflict, the variable comes from UCDP/PRIO data set (v3.0).[30] Some scholars argue that the costs of civil wars greatly affect civil war dynamics.[31] As the burdens of war become intense for the society at large, so do the incentives for all sides to find a solution to bring the war to an end.[32] More-intense conflicts are thus expected to shorten conflicts by encouraging negotiations among parties that are burdened by the war.

*Mountainous Terrain:* The variable denotes the presence of mountains in the region and comes from Fearon and Laitin (2003).[33] While not including rough terrain such as swamps and jungles, this variable measures the proportion of the mountainous part of the country. Small groups can retreat to rugged terrain, such as high mountains, dense forests, and other inaccessible landscapes to train, regroup, and hide from government forces, all of which provides the rebels with the means to sustain fighting.[34] Mountainous terrains thus create a safe haven for rebels, hindering the government's effort to defeat them. Even the viability of weak and small groups will be higher if they are based in the mountainous terrain. We took the log transformation of this variable to avoid skewness.

*Ethnic Fractionalization*: The variable, which is logged, captures the probability that two randomly selected individuals will be from different ethnic groups. As above, we obtained it from Fearon and Laitin (2003). It is likely that rebels might face greater challenge in mobilizing supporters in more ethnically divided societies. Their ability to offset the power of the government would be limited, making it easier for the government to either defeat the group or push for negotiations and in doing so end the war quickly.

*Rebel Support and Government Support*: These two variables show external intervention on behalf of the rebels and the government, respectively. They come from Cunningham et al. (2009)[35] and are updated for recent years. Findings in general show that external interventions on behalf of the warring parties prolong conflicts as they increase the motivation of the parties to continue the fighting. Balch-Lindsay and Enterline (2000)[36] find that interventions on the side of the government prolong conflicts, while Regan (2002)[37] finds that balanced interventions, military and economic ones, all lead to longer wars.

*Population*: Data on population come from World Bank Development Indicators;[38] civil war literature shows that highly populated countries tend to be associated with longer civil wars.[39] Not only is recruitment less likely to be a problem when combatants can choose from a large supply of potential individuals, but highly populated countries also create monitoring problems for governments, thus making it possible for rebel groups to survive. This variable is logged to avoid skewness.

## Selection Bias Test Variables

Additionally, for Tables 5.5 and 5.7 (selection bias tests), we coded *Resources* as 2 if the country had both diamonds and oil, 1 if it had only one of the two resources, and 0 otherwise. The information on resources comes from Humphreys (2005).[40] Data on domestic *Terrorist Attacks*, an ordinal variable denoting the number of terror attacks in a country, come from the Global Terrorism Database, specifically from Enders, Sandler, and Gaibulloev (2011),[41] who have decomposed the Global Terrorism Database into transnational and domestic events. Additionally, *Number of Groups* denotes the number of rebel groups fighting in each conflict and comes from UCDP/PRIO Armed Conflict data set (v4.0).

## Diagnostics, Time-Varying Covariates, and Robustness Tests

First we checked the correlations between the variables in both data sets. In our data that include minor and major conflicts, two variables, namely *Population* and *Proportion of Forces*, were slightly correlated. The correlation between the two variables was 0.56 for minor conflicts and 0.60 for major conflicts. We dropped *Proportion of Forces* in relevant models. The results remained the same.

Second, one might claim that an intervention in earlier stages of the conflict might have a different impact from the one that occurs later. For example, the risk of conflict termination when a government hires a publicly owned PMSC in the first year of conflict may not be the same if it were to hire a PMSC in the conflict's fifth year. It may be, for example, that PMSC performance may get better through time as PMSCs gain more experience. PMSCs may be better acquainted with the local dynamics of the war and may find more effective ways to overcome the challenges as they receive more information. In contrast to this, it may also be the case that early intervention can help defeat the rebels when they are at their weakest, long before they acquire internal and external resources and recruits to help sustain their survival.[42] To examine if timing of intervention matters, one can interact the variable of interest with time $t$. These variables are referred to as time-varying covariates in duration models. However, the tests we conducted for violation of proportionality assumption in Cox models did not detect the need for incorporating the variables that capture time for PMSC intervention. This means that the timing of PMSC intervention is not important for estimating the duration of war.

In all models, we analyzed the residuals, which are estimated as a function of ln time; if proportional assumptions hold, these residuals should be a random walk unrelated to survival time. Using Schoenfeld and scaled Schoenfeld residuals, we detected that there were mainly three variables, *Support Rebels, Mountainous Terrain,* and *Polity,* for which this assumption was specifically violated in some models. We applied time-varying covariates for these variables in relevant models.

We also calculated concordance probability, the probability that predictions and outcomes are concordant. This should help us judge the success of our model predictions. The value of Harrell's C, the proportion of all usable subject pairs in which the predictions and outcomes are concordant, was 0.71 for Model 3 in Table 5.1. This indicates that we can correctly predict survival times for pairs of conflicts 71% of the time on the basis of measurements of all the variables included in the model. We also calculated the value of Harrell's C for Model 3 in Table 5.4 as 0.76. This indicates that we can correctly predict survival times for pairs of conflicts 76% of the time on the basis of measurements of all the variables included in the model.[43]

# 6

# Under Pressure

## The Promise and Limitation of Competition

For years, the forces of the free market have fascinated the public, policymakers, and academic community. Whether it is about exalting the power of the market to benefit consumers by pushing companies to innovate and lower prices or dissecting its negative side effects linked to corporations' preoccupation with profit at the expense of ethical considerations, discussion about the free market continues to generate controversy. This book seeks to uncover how two specific market mechanisms, the level of global or industry-wide competition in the private military and security industry and local competition in a specific conflict zone, have affected international PMSCs' military effectiveness and accountability in civil wars from 1990 to 2008. In most cases of international PMSCs' interventions into civil wars, it is the central government that hires such companies as it seeks to alter the balance of power between itself and the rebel challengers. When market forces create an incentive for PMSCs to be more accountable to their clients, these actors can improve state capacity and pave the way for a quicker termination of conflict.

## Contribution

This book builds upon growing interest in the scholarly literature on PMSCs' impact on specific conflict dynamics. Instead of resorting to generalizations that place the industry in the category of villains or heroes, existing research increasingly seeks to unpack different effects of PMSCs and conditions associated with such outcomes. Tkach (2019),[1] for example, examines how variation in contract ambiguity and intra-sector competition impact levels of insurgent attacks in Iraq, while Petersohn (2017)[2] shows that PMSCs' interventions in conflicts with weak states are linked to an increase in casualties depending on the tasks the companies perform. Dunigan

*Private Militaries and the Security Industry in Civil Wars.* Seden Akcinaroglu and Elizabeth Radziszewski, Oxford University Press (2020). © Oxford University Press. DOI: 10.1093/oso/9780197520802.001.0001.

(2011)[3] investigates how effective PMSCs are when they are co-deployed. Concentrating on Iraq, she argues that co-deployments involving troops and PMSCs result in structural and identity-related integration dilemmas that decrease effectiveness, while the opposite is true when PMSCs are either structurally integrated into the military force or operate in place of the military. Building on these findings, Fitzsimmons (2013)[4] uses several case studies to examine how normative versus realist theories fare in explaining differences in mercenaries' military effectiveness, ultimately concluding that effectiveness rises in the presence of norms that encourage creative thinking, initiative, transmission of information, group loyalty, technical expertise, and decentralized decision-making.

Overall, we contribute to this growing trend in the literature that diversifies its focus on outcomes, and to the general quantitative research on PMSCs by relying on new data and theoretical insights that improve our understanding of conditions under which these corporate actors shape the duration of civil wars. First, our study explores the duration of civil wars from 1990 to 2008. As such, we introduce a new outcome variable with a global outlook. Data on PMSCs' interventions into civil wars are limited. Aside from Petersohn's work that focuses on weak states (2017,[5] 2014[6]) and Akcinaroglu and Radziszewski's data on Africa (2013),[7] the more comprehensive data come from Avant and Neu (2019).[8] While Avant and Neu's data cover a longer time frame, 1990–2012, our data account for all cases of international PMSCs' interventions into civil wars, with limitation for Afghanistan and Iraq. For those two conflicts we have an incomplete list of international PMSCs' presence on yearly basis that is nevertheless sufficient to allow us to assess the presence or absence of competition. As such, both Afghanistan and Iraq are included in the analysis and contribute to our insights. Unlike Avant and Neu (2019),[9] we focus specifically on international PMSCs' interventions into major and minor civil wars rather than on all events involving any private security contractor. Finally, we include data on fraud and human rights abuses committed by international PMSCs' in Iraq, a case that is excluded from Avant and Neu's data. Our data note whether any PMSC that has been present in Iraq was connected to crime; the type of crime and the date when it was committed. These data have enabled us to address key questions regarding international PMSCs' role in civil wars, including: does global competition that push some companies to alter their corporate structure lead to positive developments in the area of corporate professionalism? To what extent does corporate restructuring then affect companies' contribution to shortening wars? Does local competition at the conflict level lead to the same positive effect?

Second, from a theoretical perspective we develop an argument about ways in which market pressure could generate specific dynamics that could push companies to be more militarily effective in shortening the wars and thus more accountable to the client. In Chapter 2, we examined in greater depth the variety of existing mechanisms employed to address the problem of unaccountable behavior that could be an issue for clients dealing with profit-driven entities. Improving contract clarity, for example, might help in identifying specific performance expectations. Requirements that companies provide certification of meeting best practices by signing the International Code of Conduct for Private Security Service Providers or demonstrate compliance with the ANSI/ASIS or ISO Standards show that companies are improving their level of transparency, personnel vetting, and adherence to international humanitarian law. Still, these may not be sufficient to deter companies from criminal practices that might be harder to detect in conflict zones. Market-driven competitive pressure at the local and global level might contribute to existing efforts to bridge the gap between the interests of the client and the PMSC by improving military effectiveness through improvements in monitoring of companies operating in conflict zones.

Often, analysis of military effectiveness includes interest in the expected and important elements such as PMSCs' skills and organizational culture. Our focus is on skill but even more so on previously unexplored dimension of military effectiveness in the form of corporate professionalism: commitment to the culture of integrity, as evident by limited engagement in fraud and human rights abuses. In Chapter 2 we discussed the ways in which corporate professionalism merits consideration when exploring military effectiveness in the context of civil wars. While refraining from killing civilians might not have a larger impact on shifting the balance of power in international conflicts and is rarely acknowledged as a source of power, in insurgencies the rebels' strength stems from winning the support of the locals. Therefore, lack of practices designed to reduce indiscriminate violence by PMSCs and limited willingness to take these seriously could hurt the companies' effectiveness to help the governments make substantial gains against the enemy. Similarly, fraudulent practices lead to wasted resources that otherwise might be redirected to mission-related initiatives. Billions of dollars have been lost in overbilling in Iraq as some companies charged for unqualified personnel or for ghost employees. Such problems are likely to be even greater in weaker states with limited institutional development to monitor contract abuses.

We argue that there are, in fact, two different types of competitive pressure that impact the extent to which PMSCs seriously embrace corporate professionalism and thus enhance their military effectiveness in ways that make them more accountable to the client. There is competition at the conflict level and another at the industry level. One of the problems with only looking at the increase in competitive pressure at the industry level is that while it might create an incentive for international PMSCs to highlight their improvements in corporate professionalism to the clients that might increasingly demand it and seek industry and ANSI/ASIS or ISO Standards certifications, there is still informational asymmetry between the client and the service provider regarding commitment to such practices in a conflict zone. This makes it harder for the client to determine whether a company's apparent commitment to avoid indiscriminate violence against the locals will be honored. With more providers on the ground, an increase in the local, conflict-level pressure could reduce the risk for the government. With more companies on the ground there is a possibility that rival companies could observe and report criminal behavior. This informal type of peer monitoring could put pressure on companies to honor the commitments they make to the client.

What happens, on the other hand, if local competitive pressure is missing in a given conflict? Does this imply that accountability is lost? Here, we argue, is where industry-level competition can still play a role in some contexts. As more companies enter the market with an increasingly diverse portfolio of services, there will be a need to secure contracts not only by demonstrating the skills and attractive cost package but also in responding to more diverse clients' needs. When a company responds to this growing pressure by seeking new funding and exposure by going public, it engages in self-regulation that can have a promising side effect on its behavior in conflict zones. In Chapter 4 we argued that greater transparency of publicly traded PMSCs, combined with more diverse stakeholders, puts more pressure on such companies to seriously improve their approach to corporate professionalism. Put simply, exposure to scandals carries greater risk for such companies. When uncovered, the risk could be a potentially more significant loss of capital for publicly traded PMSCs than for private ones. DynCorp's story shows that while even publicly traded PMSCs can recover from scandals, the loss can include a substantial drop in stock prices and negative growth prognosis that could further undermine shareholders' trust. In the end, companies may simply deem the financial risk so costly that they revert back to their private, less transparent corporate structure.[10] When companies choose not to change their corporate structure in response to competitive pressure at the industry level, the impact of other strategies they may pursue, such as investment in new technologies or skills, may not offer the same type of benefit toward improving accountability. Overall, the book's major theoretical addition is its exploration of how these two different mechanisms of market pressure could have a varying impact on PMSCs' behavior in conflict zones.

## Key Insights

Our findings highlight the benefits of two market dynamics, increase in local and global competition, on conflict termination for most types of civil wars. First, we find that an increase in competition at the local or conflict level among international PMSCs decreases the duration of major conflicts, wars that include conventional tactics in addition to elements of guerrilla warfare. Our results from Chapter 5 show that there is almost no difference in the odds of war termination in a conflict that has been ongoing for five years without PMSC intervention in comparison to one where the government relies on services of only one company. The value of PMSCs' interventions occurs when multiple companies assist the government, with the odds of conflict survival decreasing by nearly 40% when five companies intervene. This effect is not merely driven by accumulation of resources from multiple companies. Our findings show that even when we take into account the number of services delivered to the government, the variable that captures the number of companies present in a conflict zone is still statistically significant. As such, we are able to conclude that competitive pressure plays a unique role in improving the quality of services. This benefit is also evident for all service sectors, with the exception of intelligence and communication provision.

But there is a limitation to how much competitive pressure can help governments shift the balance of power in their favor. In minor conflicts or those where guerrilla tactics dominate, competition does not bring the same results as in major wars. This occurs because multiple companies are more likely to be hired by governments in a specific subset of wars, those that are especially hard to resolve and present the most severe challenge to the government. Conflicts with multiple insurgent groups that have access to resources, rely heavily on terrorist tactics, and have a higher proportion of forces to those of the government are more likely to have interventions by multiple companies. We thus see a selection bias whereby multiple companies are linked to longer conflicts in a subset of the most challenging conflict environments. It does not mean that competitive pressure, however, is irrelevant; rather, that the findings for the most difficult cases show that in spite of competitive pressure, these companies may still struggle with helping to terminate the war. The finding thus highlights the importance of recognizing that PMSCs' capacity to be militarily effective in the most challenging contexts is dependent on the evolving nature of conflicts and not merely on their own capabilities and potential for greater accountability. As such, it might be prudent to withhold judgment on contractors for not bringing immediate gains to the governments in such contexts and recognize that varying levels of conflict complexity may play a role in shaping the success of even the most accountable companies.

Second, in Chapter 5 we were also able to examine how the rise in global competition marked by the increase in the number and diversity of security providers has affected conflict duration. We argued that with global competition, some companies might seek access to capital to invest in their growth. An important source of capital comes from declaring an initial public offering. The benefit of companies' shift in corporate structure to publicly traded entities is promising for conflict dynamics. As publicly traded PMSCs have only intervened in minor conflicts, we can see their impact on duration in such contexts. Specifically, the presence of publicly traded PMSCs is linked to shorter conflicts, and for a war that has been ongoing for 10 years, an intervention by this type of actor could offer over 18% improvement in the odds of conflict termination in comparison to contexts with interventions by private PMSCs.

It is hard to access which of the market mechanisms, local or global competition, is more significant for conflict termination. There have been no interventions of publicly traded PMSCs in major wars from 1990 to 2008 and so we cannot derive a comparison for conflicts involving mixed conventional and guerrilla tactics with those where guerrilla tactics mostly dominate. Only post-2008 do we see the presence of publicly traded companies in major wars, specifically in Iraq and Afghanistan when the two conflicts meet the casualty threshold to fall into the category of such wars. There is also evidence that while the hardest-to-resolve conflicts, minor wars with mostly guerrilla tactics, are more likely to involve multiple PMSCs and less likely to involve publicly traded PMSCs, publicly traded PMSCs do not nec-essarily shy away from difficult conflict zones. In fact, our analysis shows that they are more likely to enter into what we would refer to as semi-challenging conflicts involving multiple insurgent groups that use terrorist tactics. As such, it would be incorrect to assume that publicly traded PMSCs' impact on war duration is minimal because they already pre-select the easiest conflicts. Nevertheless, we find that the most difficult conflict dynamics that involve the above characteristic in addition to a much stronger rebel force with ac-cess to resources are linked to interventions by multiple PMSCs and not pub-licly traded PMSCs.

Overall, the findings suggest that both market mechanisms have a posi-tive effect on conflict termination, but there is some limitation to the benefit. Mainly, local competition has a positive effect in major wars involving con-ventional tactics and some guerrilla warfare but is less significant in the most difficult conflict environments. Companies' corporate structure, a factor linked to the global-competition dynamic, can affect military effectiveness in conflicts involving less casualties and more guerrilla warfare. The positive effect is also evident in hard-to-end conflicts but not in the subset of the most complex environments, as publicly traded PMSCs may strategically avoid the most extreme conflict zones.

Third, we argued that market pressure improves PMSCs' military effec-tiveness by pushing companies to be more accountable in the area of cor-porate professionalism. Competition at the local and global level leads to improvements in ethical behavior, which plays a role in saving governments' resources and winning the locals' support so vital to shifting the balance of power more favorably toward the government. While we presented examples and quotes from interviews in Chapter 3 to illustrate that local competi-tion plays a role in international PMSCs' approach to corporate profession-alism, in Chapter 4 we were able to empirically examine whether companies that change their corporate structure in response to global competition are also more serious about embracing corporate professionalism. Specifically, our case study of fraud and human rights abuses committed by PMSCs in Iraq from 2003 to 2019 demonstrates that publicly traded PMSC are less likely to be linked to such crimes. Thus a decision to go public in response to growing competition at the industry level delivers a positive side effect. It improves public companies' corporate professionalism. We then see that such companies are more militarily effective than private PMSCs in con-tributing to conflict termination because they are less likely to alienate the locals, more likely to refrain from indiscriminate use of force, and more risk-avoidant when it comes to defrauding their clients than private companies. Undoubtedly, publicly traded PMSCs are not immune to scandals but our analysis demonstrates that when it comes to corporate professionalism as a dimension of military effectiveness such actors, on average, score better.

Finally, we can shed some light on the utility of PMSCs' interventions in comparison to joint interventions involving PMSCs and outside states or international organizations. When local competition exists, PMSCs' interventions are more promising for conflict termination than interventions in the absence of such competition but in the presence of non-PMSC interveners. The latter might involve contexts where the government works with a single company but also relies on states or international organizations for support. Overall, mixed interventions are uncommon; usually governments rely either on PMSCs or on other states' and international organizations' support. It appears that the benefit of competitive pressure comes from PMSCs' themselves and not from another state that might be monitoring private contractors in the field. Companies might be adjusting their practices based on expectations that other corporate actors might informally monitor their behavior but not another intervening state. This is quite plausible because the intervening state is not directly competing with a PMSC; in other words, PMSCs' interest is profit-oriented while the other's goal might be something entirely different and involve broader geopolitical concerns. The finding also supports Dunigan's work (2011)[11] on co-deployed missions in Iraq. Mixed interventions, involving the US military and a PMSC, decrease PMSCs' military effectiveness. Dunigan argues that problems with integration and identity-related issues make such missions challenging, while our findings show that mixed interventions may not be as effective in enhancing PMSCs' accountability as interventions involving only the same type of actors, multiple PMSCs.

## Future Focus

The private military and security market is a crowded one, and international PMSCs are only a subset of actors that provide security services to a wide range of clients, from governments, international organizations, to insurgents and private businesses. There is every reason to expect that as demand among these different clients for security provision continues, providers will continue to face some form of competitive pressure. We argued that international PMSCs' pressure comes in large part from the companies' long-term perspective of not only surviving but also thriving on the market that caters to highest-paying clients, which most often include mega consumers like the US and the UK. Such expectations bring a unique set of challenges for international PMSCs that must respond to these needs by constantly evolving to meet the clients' demands or lose a contract bid to another firm. Yet for many mercenaries and quasi-PMSCs that sometimes operate on ad hoc basis, the goal may not necessarily be to survive on the security market in the long run. Instead, such actors might seek to secure a new lucrative contract and then move on to other, unrelated businesses or cease to exist. Sabre International Security, a two-man venture set up in Iraq by a veteran of British Special Air Service and an Iraqi businessman, amassed $300 million worth of contracts to guard the embassy in Iraq but after a fallout with Iraqi prime minister, the company simply vanished.[12] Nevertheless, such actors may still compete with other quasi-security providers to secure a client. Future studies might investigate the different types of competitive pressures on private security contractors other than international PMSCs and address the outcome of their activities. For example, if such actors are less concerned about their long-term reputation, they may exhibit an organizational structure that is less focused on specific rules designed to promote accountability and limit the occurrence of fraud and humanitarian violations. This, in turn, might suggest that quasi-companies would be less militarily effective. While they might help the client secure some victories their strategies might hurt the client's overall security goals with detrimental effect on conflict termination.

Delving deeper into the interests and operations of such actors is especially relevant today, with a "return" of the mercenaries. The time period from 1990 to 2008, which is the focus in our book, coincides with the rise of PMSCs that develop a clear corporate structure. The latter will continue to remain relevant. Yet the recent use of mercenaries in places such as Yemen, Syria, and Nigeria suggests that security contractors might be operating less as corporate entities and more as loose organizations without clear hierarchical structures in place. What can we learn about the identity of such actors and how they operate? How do they differ, if at all, from mercenaries who have provided services to governments before the emergence of corporate warriors post-1990?

One hypothesis might be that today's mercenaries might be working as proxies of governments that seek to influence conflict dynamics around the globe with plausible deniability and those that want to achieve their geopolitical objectives without directly facing more conventionally superior enemies. Russia, for example, has followed the Gerasimov doctrine, named after the country's chief of the General Staff, which calls for the use of guerrilla and conventional tactics to weaken the US and its allies.[13] Enter the Wagner Group, a mysterious Russian mercenary organization with connections to the Kremlin. The Wagner Group, dubbed "little green men" in the Western press,[14] has been active in Syria's and Ukraine's civil wars, and its presence has been noted in Africa since 2018.[15] The use of Wagner Group and other tactics of irregular warfare including cyberattacks has been a significant part of Russia's broader strategy of spreading chaos, in essence a move designed either to keep conflicts ongoing[16] or to bolster the presence of anti-Western regimes. This trend is likely to continue; in 2018 a company called Patriot has popped up in Russia and appears to model its activities after Wagner Group in a way that suggests the growing importance of such organizations in Russia's foreign policy.[17]

While the use of mercenaries as proxies of the state is not new, Britain, for example, has relied on mercenary outfits to assist friendly governments during the Cold War when it did not want to be openly linked to such activities,[18] Russian mercenary organizations are more closely connected to the state apparatus and ideology. Patriotism is the motto for Wagner fighters, as noted by one of its commanders: "Even if you are 10,000 kilometers from home, you are still fighting for the motherland."[19] They also operate on a global scale through much more informal networks between buyers and sellers and with limited domestic oversight.[20] This phenomenon has the following implications. First, it suggests that to derive meaningful conclusions about such actors' impact on global security, their activities must be analyzed through the interests of the states they are informally connected to. Second, as such companies operate with limited oversight and are well connected to the state apparatus, even if they deliver suboptimal service to the client, they may still survive on the market. This paints a rather grim picture, as it suggests that with limited market pressure and oversight, such organizations might have a negative impact on conflict termination. For example, they might provide security services to the client without consideration for civilian casualties or work for both sides as the Wagner Group has done by delivering services simultaneously to the government and the rebels in Central African Republic.[21]

Interestingly, it might be of great value to international PMSCs to understand who these new players are, as the former might have to be prepared to tackle the challenge stemming from mercenaries if both types of actors inter-vene in the same conflict but on different sides. The challenge for researchers will lie in gathering relevant insights and data on groups that thrive in the underground and want to remain hidden from the public's view.

As the case of Wagner Group shows, unpacking the relationship between private military and security contractors' presence in conflict zones cannot be entirely isolated from the broader conflict environment. The ability of PMSCs to improve the security situation in insurgencies can be better under-stood when their presence is studied in connection with the social, political, and economic conditions they face on the ground. Even if companies adhere to corporate professionalism and provide promised services and do so in a way that limits civilian abuses, their level of military effectiveness, while gen-erally better than in the absence of such practices, might still vary depending on whether the company operates in areas with different levels of population density, ethnic diversity, and influx of displaced persons. Our findings sup-port this idea; even when multiple companies are present in the subset of the most difficult conflict zones they may not be sufficient to help with quick ter-mination of conflict.

To better understand ways in which complex conflict dynamics affect mil-itary effectiveness, data disaggregation might be helpful. In places such as Iraq there could be significant variation in cultural sensitivities that could also impact companies' military effectiveness. Thus PMSCs' engagement in a Diyala Governorate, considered the most diverse area in the country with different ethnic and religious groups existing in the same space and with growing population of internally displaced persons,[22] might generate worse results than similar interventions in other governorates because the conflict environment there might be more conducive to civilian clashes in the first place. Such selection biases need to be considered to avoid either inflating the positive impact of PMSCs or undervaluing their benefit. Disaggregation can also help uncover how failures in one area can have a negative spillover effect into other area, which in turn could aggregate to overall mission failure. One place to start would be Iraq and Afghanistan where data on US-hired PMSCs' presence at the local, provincial level is available through the Freedom of Information Act. In fact, research by Tkach (2019)[23] is the first to examine governorate-level data on PMSCs in Iraq from 2004 to 2008, with a focus on how intra-service competition and contract clarity affect severity of violence.

Just as variation in cultural sensitivities might impact conflict severity and duration, so does the evolution in rebel strategies. Insurgents adjust their strategies in response to governments' and PMSCs' strategies. In Iraq, the use of improvised explosive devices became more sophisticated over the years and accounted for an increase in military and security personnel casualties as fighters learned to better position roadside bombs and gained new skills to improve remote denotation.[24] Boko Haram's sophistication in Nigeria was evident when the organization turned to drones and mercenaries in 2018, a move that caught the Nigerian military by surprise even after that government supposedly relied on former Executive Outcomes men to deal with the problem.[25] If the boost to Nigerian military was temporary, it may be because the government and its security providers underestimated the way in which rebel organizations evolve. Evolution in novel and creative tactics is also present among terrorist organizations that innovate to gain a competitive advantage and desired outcome.[26] Such dynamics introduce an element of complexity into the conflict that can negatively affect PMSCs' ability to fulfill contractual obligations in the optimal way. Studies could thus analyze longitudinal data in Iraq and Afghanistan to determine the extent to which limited military effectiveness of PMSCs' operating in a specific local context might be attributed more to the evolving conflict complexity than to a deliberate lack of accountability, as is sometimes assumed in the scholarly and mainstream literatures.

## Policy Implications

While market mechanisms can improve PMSCs' accountability, the effect is conditioned on two main factors. First, policymakers must display commitment to developing and upholding criteria for competitive contract awards that will set the stage for more accountable performance by companies in a conflict zone. It means that smaller contracts diffused among several companies within the same sector might encourage the kind of competitive pressure that will push companies to refrain from fraud and human rights abuses or risk exposure from competitors and bad publicity that could curtail their ability to receive new contracts. While awarding contracts to publicly traded PMSCs based exclusively on companies' corporate structure would be considered discriminatory, contract specialists can nevertheless create criteria for transparency that might implicitly favor this type of actor. In the case of the US, for example, the DoD allows contracting officials to develop non-cost factors for evaluating companies' potential to accomplish the job.[27] It means that while contracts are based on lowest-price, technically acceptable standards, there is also room to address accountability issues through serious consideration of relevant and new non-cost factors. Contracting officials might, for example, develop non-cost factors that could include not only companies' disclosure of risks and past performances in the area of human rights protection, but also the ratio of positive to negative media coverage in the past year and media's investigations of alleged criminal activities. Our findings show that publicly traded PMSCs would be more likely to score higher on such dimensions. With such companies on the job, especially in areas where contact with the locals is expected to be high, there is greater likelihood of respect for human rights, which ultimately can increase the success of the overall mission in a conflict zone. Some of these criteria extend beyond the ANSI/ASIS and ISO Standards' requirement on personnel vetting, proper qualifications, and mechanisms to ensure compliance with international humanitarian laws. And while they are already likely to be the focus of public companies due to their greater reputational costs, they might also push private PMSCs to be more serious about embracing a record with limited scandals if they seek to compete with public companies in this area.

While developing specific contract requirements that set the stage for market dynamics to work in a positive way is one way to improve accountability, ultimately it is up to the client to display willingness to do so. It also means that clients must punish companies that fail to be accountable. Competitive pressure works when companies expect that revelations of abuses will adversely affect their profitability. As such, clients must issue costly signals to convey their seriousness in upholding the standard. In practice, however, this has not always been the case, as we highlighted in our discussion on market limitations in Chapter 4. At different stages in the Iraq war there was inconsistency in exercising punishment for companies' failure in accountability. In 2005, CACI International and Titan both received new contracts despite reports that their employees were involved in the Abu Ghraib prisoner abuse scandal.[28] While companies need not be barred from bidding for a contract for poor performance forever, nevertheless there should be limitations in place that forbid companies from bidding for a set time period, with specific requirements for companies to document steps they have undertaken to prevent another abuse from occurring. Even more important, contracting departments should develop a common policy of contract non-renewal across all agencies to impose a heavy cost on the company in the midst of scandal revelations. Even though a company may feel a dent in its profits when denied a new contract from the Department of State, shareholders may still have confidence in the company if a new contract comes from the DoD. Lastly, governments must avoid no-bid contracts. In the US the occurrence of such contracts was more common in the early stages of the Iraq war, and declined after a March 4, 2010, memorandum from President Barack Obama instructed the Office of Management and Budget to "maximize the use of full and open competition" in the awarding of federal contracts.[29] Still, even after such a decision, the practice has not completely gone away.[30] It is also likely that in the contexts outside of the US, particularly in new democracies, no-bid contracts are more common. Such practices should be avoided for market dynamics to work in a way that strengthens accountability.

Second, policymakers might consider broadening the responsibilities they assign to intervening PMSCs that include greater engagement with the hearts-and-minds approach that responds to evolving conflict complexities. While this is undoubtedly controversial as it may require more contribution from PMSCs into strategy development, it could benefit particularly those states that struggle in this area. In the 1990s there were concerns about PMSCs taking over the state's monopoly on the use of force. More recently, Erik Prince's proposal to embed private contractors as "mentors" into Afghan forces, create a private air force to compensate for the government's air deficiencies, and appoint a viceroy, a federal official who would oversee the contractors' mission and report to the US president, has been dismissed by the Afghan government.[31] While Prince's proposal carries a number of significant problems and in essence lacks novel solutions in strategy development that would build popular support for the Afghan forces, nevertheless it speaks to a much bigger and ongoing question pertaining to the evolution of contractors' presence in conflict zones.

Our study shows that in the most complex conflicts—those where rebel forces enjoy access to resources, have a higher proportion of forces to those of the government, and are split into several distinct groups—the presence of multiple PMSCs may not be sufficient to bring about a quicker termination of war despite potential for competitive pressure to limit fraud and human rights abuses, factors that contribute to greater military effectiveness in most other conflict milieus. Such complex conflict scenarios may require application of new, more in-depth COIN practices designed to secure the support of the population while concurrently cutting off rebels' access to material resources that fund the war. In such extreme contexts, governments should be, at the very least, open to proposals from multiple PMSCs, though not a single one, to develop a diverse and new strategy for multiple companies to assist the government not by providing more of the same but in highlighting their role in expanding the scope of population-centric COIN. With local competitive pressure contributing to companies' adherence to international humanitarian law, there is greater chance that new ideas about population-focused efforts could bring improvements. PMSCs could also help militaries develop the necessary skills to combine the latter more effectively with offensive operations.

At the same time, governments must also recognize that addressing the global dimension of many insurgencies may be necessary to limit the source of rebel funding that keeps these conflicts ongoing. Managing this conundrum may extend beyond the scope of international PMSCs' evolving support in population-centric missions. In essence, these cases require governments to embrace a multidimensional approach involving the support of PMSCs and cooperation from other states in curtailing regional and global illegal trade in resources that help fund the insurgents.

It is also worth noting that the best strategy for managing conflicts based predominantly on guerrilla warfare would not necessarily involve direct interventions from PMSCs *and* other states. Our findings show that this approach would not bring meaningful gains. If governments insist on receiving assistance from an outside state and a PMSC concurrently, they should establish training programs for these two types of intervening actors not only to clarify mission coordination but, most important, to agree on how to handle incompatible interests that could surface in strategy development. For example, an international PMSC that happens to be a publicly traded company is likely to display risk-aversion with respect to civilian casualties due to reputational costs. An intervening state, however, might seek to rally its own public behind an aggressive, diversionary policy to score points at home, and could thus push for a quick offensive with limited regard for civilian casualties. Such a strategy frequently backfires in the long term and prolongs the war for the state experiencing the conflict. Contracting governments must be aware of intervening actors' competing interests and ensure that they are addressed early in the mission.

Finally, companies must expect that their competitors who witness PMSCs' waste and human rights abuses will, in fact, report misgivings. The logic of deterrence suggests that as long as such possibility exists, companies might exercise caution. Governments might strengthen this perception by clearly communicating new efforts to protect and reward whistleblowers. In the US context, the 2013 National Defense Authorization Act offers protection to contractors and subcontractors when they report fraud and abuses, while under the False Claims Act if an employee faces retaliation from an employer for considering reporting fraud, the employee is entitled to double damages. However, these initiatives could be improved by extending protection and rewards to whistleblowers outside the US.[32] An employee of a British PMSC, for example, might be encouraged to report criminal behavior witnessed by an employee from an American PMSC to a contracting agency of the latter if whistleblowing protection covers reports in a conflict zone by a foreign national. For weak states with limited institutional development, the industry itself might step in to help lead the effort on educating governments on the benefits of such programs. Such steps would further signal international PMSCs' effort to strive for accountability as an industry and highlight to clients who are new to the international PMSC market how this subset of corporations differs from ad hoc companies or mercenaries in promoting accountability as a component of military effectiveness. Similarly, individual PMSCs should invest more effort into rewarding their own employees for reporting abuses committed by employees from other companies by establishing formal programs that deal specifically with such cases and communicating these initiatives publicly to generate industry-wide awareness of their existence. This, in turn, strengthens companies' expectations that their abuses could be reported and exploited by the competitor.

# Notes

## Chapter 1

1. "Verbatim," *Time*, September 27, 2010, 3.
2. Peter W. Singer, *Corporate Warriors: The Rise of the Privatized Military Industry* (Ithaca, NY: Cornell University Press, 2003), 88–100.
3. Data taken from UCDP/PRIO Armed Conflict Dataset version 19.1. See, Therese Pettersson, Stina Hogbladh, and Magnus Oberg, "Organized Violence, 1989-2018 and Peace Agreements," *Journal of Peace Research* 56, no. 4 (2019): 589–603; Nils Petter Gleditsch et al., "Armed Conflict 1946–2001: A New Dataset," *Journal of Peace Research* 39, no. 5 (2002): 625–637.
4. Alina Rocha Menocal, Verna Fritz, and Lisa Rakner, "Hybrid Regimes and the Challenges of Deepening and Sustaining Democracy in Developing Countries," *South African Journal of International Affairs* 15, no. 1 (2008): 29–40.
5. "Global Conflict Trends," Center for Systemic Peace, http://www.systemicpeace.org/conflict.htm, accessed May 15, 2013.
6. Patrick Regan, "Third Party Interventions and the Duration of Intrastate Conflicts," *Journal of Conflict Resolution* 46, no. 1 (2002): 55–73; Ibrahim A. Elbadawi and Nicholas Sambanis, "External Interventions and the Duration of Civil Wars," World Bank Working Paper, September 1, 2000, 1–18, http://documents.worldbank.org/curated/en/760801468766521682/External-interventions-and-the-duration-of-civil-wars.
7. Dan Miodownik, Oren Barak, Maayan Mor, and Omer Yair, "Introduction," in *Nonstate Actors in Intrastate Conflicts*, ed. Dan Miodownik and Oren Barak (Philadelphia: University of Pennsylvania Press, 2014), 1–12; Nava Lowenheim, "Turkey's Dual Problem: Between Armenia and the Armenian Diaspora," in *Nonstate Actors in Intrastate Conflicts*, ed. Dan Miodownik and Oren Barak (Philadelphia: University of Pennsylvania Press, 2014), 116; Jolle Demmers, "Diaspora and Conflict: Locality, Long-Distance Nationalism, and Delocalisation of Conflict Dynamics," *Javnost—The Public* 9, no. 1 (2002): 85–96.
8. These include research on the impact of PMSCs' interventions on behalf of the government and the rebels on the duration of civil wars in Africa by Seden Akcinaroglu and Elizabeth Radziszewski, "Private Military Companies, Opportunities, and Termination of Civil Wars in Africa," *Journal of Conflict Resolution* 57, no. 5 (2013): 795–821; the impact of mercenaries and PMSCs' on the severity of civil wars by Ulrich Petersohn, "The Impact of Mercenaries and Private Military and Security Companies on Civil War Severity between 1946 and 2002," *International Interactions* 40, no. 2 (2014): 191–215 and by Ulrich Petersohn, "Private Military and Security Companies (PMSCs), Military Effectiveness, and Conflict Severity in Weak States,

1990–2007," *Journal of Conflict Resolution* 61, no. 5 (2017): 1046–1072; four case studies of PMSCs' and mercenaries' effectiveness in asymmetric civil wars by Scott Fitzsimmons, *Mercenaries in Asymmetric Conflicts* (New York: Cambridge University Press, 2003), 47–273; the connection between PMSCs' contract structure, intra-sector competition, and violence in Iraq's governorates by Benjamin Tkach, "Private Military and Security Companies, Contract Structure, Market Competition, and Violence in Iraq," *Conflict Management and Peace Science* 36, no. 3 (2019): 291–311; and the relationship between PMSCs' involvement in specific events for the government and the rebels and the duration of civil wars in Africa (1990–2008) examined by Deborah Avant and Kara Kingma Neu, "The Private Security Events Database," *Journal of Conflict Resolution* 63, no. 8 (2019): 1986–2006.

9. Anna Leander, "The Power to Construct International Security: On the Significance of Private Military Companies," *Millennium: Journal of International Studies* 33, no. 3 (2005): 803–825; James Cockayne, "Make or Buy? Principal-Agent Theory and the Regulation of Private Military Companies," in *From Mercenaries to Market: The Rise and Regulation of Private Military Companies*, ed. Simon Chesterman and Chia Lehnardt (New York: Oxford University Press, 2007), 196–217.

10. See, for example, Angela McIntyre and Taya Weiss, "Weak Governments in Search of Strength: Africa's Experience of Mercenaries and Private Military Companies," in *From Mercenaries to Market: The Rise and Regulation of Private Military Companies*, ed. Simon Chesterman and Chia Lehnardt (New York: Oxford University Press, 2007), 67–82; Deborah Avant, *The Market for Force: The Consequences of Privatizing Security* (New York: Oxford University Press, 2005), 143–147; Laura Dickinson, "Contract as a Tool for Regulating Private Military Companies," in *From Mercenaries to Market: The Rise and Regulation of Private Military Companies*, ed. Simon Chesterman and Chia Lehnardt (New York: Oxford University Press, 2007), 217–239.

11. See, for example, Herbert Howe, "Private Security Forces and African Stability: The Case of Executive Outcomes," *Journal of Modern African Studies* 36, no. 2 (1998): 307–331; Dave Whyte, "Lethal Regulation: State Corporate Crime and the United Kingdom Government's New Mercenaries," *Journal of Law and Society* 30, no. 4 (2003): 575–600.

12. See, for example, Leander, "The Power to Construct International Security," 803–824; Abdel-Fatau Musah and J. 'Kayode Fayemi, "Africa in Search of Security: Mercenaries and Conflicts—An Overview," in *Mercenaries: An African Security Dilemma*, ed. Abdel-Fatau Musah and J. 'Kayode Fayemi (Sterling, VA: Pluto Press, 2000), 47.

13. Singer, *Corporate Warriors*, viii.

14. Avant, *The Market for Force*, 221.

15. Jutta Joachim, Marlene Martin, Henriette Lange, Andrea Schneiker, and Magnus Dau, "Twittering for Talent: Private Military and Security Companies between Business and Military Branding," *Contemporary Security Policy* 39, no. 2 (2018): 298–316.

16. Scott Fitzsimmons, *Private Security Companies during the Iraq War: Military Performance and the Use of Deadly Force* (New York: Routledge, 2016), 213.

17. Olivia Allison, "Informal but Diverse: The Market for Exported Force from Russia and Ukraine," in *The Market for Force: Privatization of Security across World Regions*,

ed. Molly Dunigan and Ulrich Petersohn (Philadelphia: University of Pennsylvania Press, 2015), 92.

18. Kevin O'Brien, "What Should and What Should Not Be Regulated," in *From Mercenaries to Market: The Rise and Regulation of Private Military Companies,* ed. Simon Chesterman and Chia Lehnardt (New York: Oxford University Press, 2007), 29–49.

19. Joanna Spear, *Market Forces: The Political Economy of Private Military Companies* (Oslo: FAFO, 2000), 41–47, https://www.fafo.no/media/com_netsukii/531.pdf.

20. Avant, *The Market for Force*, 221.

21. Spear, *Market Forces*, 45.

22. Cockayne, "Make or Buy?," 196.

23. Jose Gomez del Prado, "The Privatization of War: Mercenaries, Private Military and Security Companies (PMSC)," Global Research, April 9, 2016, https://www.globalresearch.ca/the-privatization-of-war-mercenaries-private-military-and-security-companies-pmsc/21826; Singer, *Corporate Warriors*, 45–47.

24. Peter W. Singer, "Corporate Warriors: The Rise of the Privatized Military Industry and Its Ramifications for International Security," *International Security* 26, no. 3 (2001–2002): 186–220.

25. Kateri Carmola, "PMSCs and Risk in Counterinsurgency Warfare," in *Contractors and Wars: The Transformation of US Expeditionary Operations*, ed. Christopher Kinsey and Malcolm Hugh Patterson (Stanford, CA: Stanford University Press, 2012), 143.

26. Edward Luttwak, "Dead End: Counterinsurgency Warfare as Military Malpractice," *Harper's Magazine*, February 2007, https://harpers.org/archive/2007/02/dead-end/.

27. Avant, *The Market for Force*, 221; Singer, *Corporate Warriors*, 45–46.

28. Eugenio Cusumano, "Regulating Private Military and Security Companies: A Multifaceted and Multilayered Approach," EUI Working Paper, Academy of European Law, 2009, 10.

29. McIntyre and Weiss, "Weak Governments," 67–82.

30. Ibid.

31. Charles Fombrun, *Reputation: Realizing Value from the Corporate Image* (Boston: Harvard Business School Press, 1996), 37.

32. Ekaterian Balabanova, *The Media and Human Rights: The Cosmopolitan Promise* (New York: Routledge, 2015), 1.

33. Dylan Balch-Lindsay and Andrew J. Enterline, "Killing Time: The World Politics of Civil War Duration, 1820–1992," *International Studies Quarterly* 44, no. 4 (2000): 615–642.

34. Seden Akcinaroglu and Elizabeth Radziszewski, "Expectations, Rivalries, and Civil War Duration," *International Interactions* 31, no. 4 (2005): 349–374.

35. International Committee of the Red Cross and Swiss FDFA, "The Montreux Document: On Pertinent International Legal Obligations and Good Practices for States Related to Operations of Private Military and Security Companies during Armed Conflict," September 17, 2008, https://www.icrc.org/en/doc/assets/files/other/icrc_002_0996.pdf, 7–27.

36. "International Code of Conduct for Private Security Service Providers," International Code of Conduct Association, November 9, 2010, https://www.icoca.ch/sites/all/themes/icoca/assets/icoc_english3.pdf, 1–15.

37. ANSI/ASIS International, "Management System for Quality of Private Security Company Operations—Requirements with Guidance," March 5, 2012, https://www.acq.osd.mil/log/PS/.psc.html/7_Management_System_for_Quality.pdf, 1–81.

38. ISO (the International Organization for Standardization), "Management System for Private Security Operations—Requirements with Guidance for Use," September 2015, https://www.iso.org/standard/63380.html, 1–98.

39. McIntyre and Weiss, "Weak Governments," 67–82; Singer, *Corporate Warriors*, 45–47.

40. Five large-N studies that are the exception to this are Akcinaroglu and Radziszewski, "Private Military Companies," 795–821; Petersohn, "The Impact of Mercenaries," 191–215; Petersohn, "Private Military and Security Companies," 1046–1072; Tkach, "Private Military and Security Companies, Contract Structure," 291–311; and Avant and Neu, "The Private Security Events," 1986–2006.

41. Petersohn, "Private Military and Security Companies," 1046–1072.

42. Tkach, "Private Military and Security Companies, Contract Structure," 291–311.

43. Since scholars use different casualty thresholds for civil wars, our data on PMSCs' interventions capture all interventions that meet the UCDP/PRIO Armed Conflict dataset's definition of civil wars and the definition used by the Correlates of War (COW) dataset. COW's definition of a civil war includes militarized conflicts with at least 1,000 battle-related deaths and effective resistance on both sides, while the UCDP/PRIO dataset incorporates lower-intensity conflicts based on the 25 battle-deaths threshold.

44. PMSCs' interventions are noted for all conflicts that have been ongoing as of 1990, the starting point of our analysis.

45. British Foreign and Common Wealth Office, "Private Military Companies: Options for Regulation 2001–02," February 12, 2002, https://assets.publishing.service.gov.uk/government/uploads/system/uploads/attachment_data/file/228598/0577.pdf, 28–38.

46. Akcinaroglu and Radziszewski, "Private Military Companies," 795–821.

47. "Private Security Database," Data on Armed Conflict and Security, https://www.conflict-data.org/psd/index.html, accessed May 5, 2019.

48. Tkach, "Private Military and Security Companies, Contract Structure," 291–311.

49. Avant and Neu, "The Private Security Events," 1986–2006.

50. For Iraq and Afghanistan we have an incomplete list of PMSCs' presence. This incomplete list nevertheless allows us to determine that for every year during the time span of our data there was more than one PMSC present in both contexts and at least one PMSC with a publicly traded corporate structure. As such, we include Iraq and Afghanistan in our empirical analysis although we list only a sample of PMSCs' presence in our tables for companies' interventions. A full list of companies' presence in Iraq (2003-2019) is noted in Chapter 4 but it does not include specific years of intervention for each company.

51. Heidi Peters and Sofia Plagakis, "Department of Defense Contractor and Troop Levels in Afghanistan and Iraq: 2007–2018," Congressional Research Service, May 10, 2019, https://fas.org/sgp/crs/natsec/R44116.pdf, 15.

52. Micah Zenko, "Mercenaries Are the Silent Majority of Obama's Military," *Foreign Policy*, May 18, 2016, https://foreignpolicy.com/2016/05/18/private-contractors-are-the-silent-majority-of-obamas-military-mercenaries-iraq-afghanistan/.

53. Peters and Plagakis, "Department of Defense," 15.

54. Renée de Nevers, "Private Military Contractors and Changing Norms for the Laws of Armed Conflict," in *New Battlefields/Old Laws: Critical Debates on Asymmetric Warfare*, ed. William C. Banks (New York: Columbia University Press, 2011), 69–74.

55. Abisola Olasupo, "Nigeria: How Buhari Stopped Us from Fighting Boko Haram—South African Mercenary," *The Guardian*, November 26, 2018, https://allafrica.com/stories/201811270024.html.

56. Ian Bruce, "Charity Urges Westminster to Regulate UK Mercenary Firms," *The Herald*, February 13, 2008, https://www.heraldscotland.com/news/12749068.charity-urges-westminster-to-regulate-uk-mercenary-firms/.

57. Luke Harding and Jason Burke, "Leaked Documents Reveal Russian Effort to Exert Influence in Africa," *The Guardian*, June 11, 2019, https://www.theguardian.com/world/2019/jun/11/leaked-documents-reveal-russian-effort-to-exert-influence-in-africa.

58. An example of this is CACI's 2019 contract to develop and deploy new technology intelligence systems for the US Army. See "CACI Wins $415 Million Contract to Develop and Deploy Intelligence Systems for U.S. Army," Business Wire, accessed July 1, 2019, https://www.businesswire.com/news/home/20190509005154/en/CACI-Wins-415-Million-Contract-Develop-Deploy.

59. Christopher Mayer, "Private Security, Military Contractors," interview by the author, Elizabeth Radziszewski, January 19, 2019.

60. "Lead Analyst Says: Private Military Security Services Market Set to Grow to $420 Bn by 2029," Visiongain, March 13, 2019, https://www.globenewswire.com/news-release/2019/03/13/1752698/0/en/Lead-analyst-says-Private-military-security-services-market-set-to-grow-to-420-bn-by-2029.html.

61. See, for example, Beatrice Hibou, "From Privatizing the Economy to Privatizing the State: An Analysis of the Continual Formation of the State," in *Privatizing the State*, ed. Beatrice Hibou (New York: Columbia University Press, 2004), 4–5; Avant, *Market for Force*, 45–57; Oliver Williamson, "Public and Private Bureaucracies: A Transaction Cost Economics Perspective," *Journal of Law, Economics, and Organization* 15, no. 1 (1999): 306–342.

62. Avant, *The Market for Force*, 154.

## Chapter 2

1. "International Convention against the Recruitment, Use, Financing and Training of Mercenaries," United Nations Human Rights Office of the High Commissioner,

October 20, 2001, https://www.ohchr.org/EN/ProfessionalInterest/Pages/Mercenaries.aspx.

2. While scholars such as Singer, *Corporate Warriors*, 53, and Sean McFate, *The Modern Mercenary* (New York: Oxford University Press, 2014), 43–44, attribute changes in military reforms after the end of the Cold War to greater reliance on PMSCs, Elke Krahmann, *States, Citizens and the Privatization of Security* (New York: Cambridge University Press, 2010), 4–5, 84–194, brings a new dimension to the argument by focusing on the role of ideology in decisions to outsource security to private companies. She argues that outsourcing has been more significant in the US and UK contexts than in Germany because of the states' ideological differences. The UK and US governments' Neoliberal discourse places more emphasis on market competition and consumer choices, while Germany's Republican model stresses the importance of community. Consequently, Germany has been reluctant to rely on PMSCs while the UK and US have been more sympathetic toward the privatization of security and creating an opportunity for the PMSC market to flourish.

3. Former Blackwater Employee, "Private Security Accountability," interview by the author, Elizabeth Radziszewski, August 26, 2012.

4. Don Mayer, "Peaceful Warriors: Private Military Security Companies and the Quest for Stable Societies," *Journal of Business Ethics* 89, no. 4 (2009): 387–401.

5. Editorial Board, "Runaway Spending on War Contractors," *New York Times*, September 17, 2011, https://www.nytimes.com/2011/09/18/opinion/sunday/runaway-spending-on-war-contractors.html..

6. According to Fabien Mathieu and Nick Dearden, "Corporate Mercenaries: The Threat of Private Military and Security Companies," *Review of African Political Economy* 34, no. 114 (2007): 751, the industry's lobbying efforts to limit the interference of the US and UK governments in the conduct of the industry's affairs in conflict zones have contributed to this problem. In 2001, for example, the leading private military and security companies spent more than $32 million on lobbying and donated $12 million to political campaigns.

7. Molly Dunigan, *Victory for Hire: Private Security Companies' Impact on Military Effectiveness* (Stanford, CA: Stanford University Press, 2011), 87.

8. Robert J. Griffiths, *U.S. Security Cooperation with Africa: Political and Policy Challenges* (New York: Routledge, 2016), 18–21; Anna Leander, "The Market for Force and Public Security: The Destabilizing Consequences of Private Military Companies," *Journal of Peace Research* 42, no. 5 (2005): 614.

9. Jake Sherman, "The Markets for Force in Afghanistan," in *The Market for Force: Privatization of Security across World Regions*, ed. Molly Dunigan and Ulrich Petersohn (Philadelphia: University of Pennsylvania Press, 2015), 108–112.

10. Dunigan, *Victory for Hire*, 78–82.

11. Ibid., 54–63.

12. Petersohn, "Private Military and Security Companies," 1061.

13. Tkach, "Private Military and Security Companies, Contract Structure," 1.

14. Paul Richards, "West-African Warscapes: War as Smoke and Mirrors: Sierra Leone, 1991–92, 1994–95, 1995–96," *Anthropological Quarterly* 78, no. 2 (2005), 377–402.

15. Singer, *Corporate Warriors*, 166–167.

16. Leander, "The Market for Force and Public Security," 615–616.

17. Ibid., 617.

18. Income for the private security and military industry totaled approximately $100 billion in 2004, according to Mathieu and Dearden, "Corporate Mercenaries," 745. One of the largest security firms, G4S, reported over a 5% increase in earnings from 2016 to 2017. See "G4S 2017 Full Year Results," G4S, March 8, 2018, https://www.g4s.com/-/media/g4s/corporate/files/investor-relations/2017/prelim2017fullyearresults08032018.ashx?la=en&hash=0DD149ECC59490855C37772F70BA6A11, 1. DynCorp reported profit of $31 million in 2017 in comparison to a $56 million loss in 2016. See Marjorie Censer, "After Refocusing, DynCorp International Sees Profits Rise, Weighs Acquisitions," *Inside Defense*, May 29, 2018, https://insidedefense.com/share/196249.

19. Whyte, "Lethal Regulation," 583.

20. Mathieu and Dearden, "Corporate Mercenaries," 746.

21. Doug Brooks, "Write a Cheque, End a War: Using Private Military Companies to End African Conflicts?," *Conflict Trends 2000/01*, December 31, 1999, https://www.accord.org.za/publication/conflict-trends-2000-1/, 33; Kevin O'Brien, "Private Military Companies and African Security 1990–98," in *Mercenaries: An African Security Dilemma*, ed. Abdel-Fatau Musah and J. 'Kayode Fayemi (Sterling, VA: Pluto Press, 2000), 49–60, 70–71.

22. David Shearer, "Private Military Force and Challenges for the Future," *Cambridge Review of International Affairs* 13, no.1 (1999): 80–94.

23. Whyte, "Lethal Regulation," 587.

24. Doug Brooks, "Messiahs or Mercenaries? The Future of International Private Military Services," *International Peacekeeping* 7, no. 4 (2000): 129–144.

25. For a discussion of PMSCs' current role in UN peace operations, see, for example, Sean McFate, *The Modern Mercenary: Private Armies and What They Mean for World Order* (New York: Oxford University Press, 2014), 101–131; Christopher Spearin, "Since You Left: United Nations Peace Support, Private Military and Security Companies, and Canada," *International Journal: Canada's Journal of Global Policy Analysis* 73, no. 1 (2018): 68–84.

26. International Committee of the Red Cross and Swiss FDFA, "The Montreux Document," 7–27.

27. Ibid., 14.

28. Ibid., 19–20, 24, 27.

29. ANSI/ASIS, "Management System," xvii.

30. Erik Daniel Erikson, "Meeting ISO 18788 Criteria," International Foundation for Protection Officers, June 2018, http://www.ifpo.org/wp-content/uploads/2018/06/18788_EDE_article.doc.

31. Kevin O'Brien, "What Should," 29–49.

32. Marina Caparini, "Domestic Regulation: Licensing Regimes for the Export of Military Goods and Services," in *From Mercenaries to Market: The Rise and Regulation of Private Military Companies*, ed. Simon Chesterman and Chia Lehnardt (New York: Oxford University Press, 2007), 158–181.

33. Steve Fainaru, "Where Military Rules Don't Apply: Blackwater's Security Force in Iraq Given Wide Latitude by State Dept," *Washington Post*, September 20, 2007, https://www.washingtonpost.com/wp-dyn/content/article/2007/09/19/AR2007091902503_4.html?sid=ST2007092000478.

34. Caparini, "Domestic Regulation," 158–181.

35. Christopher Mayer, "DoD Monitoring," interview by the author, Elizabeth Radziszewski, August 5, 2019. Following the establishment of the Reconstruction Operations Center, the DoD also created Contractor Operational Cells to further improve coordination between the military and PMSCs in areas where PMSCs are operating. See, for example, Christopher Kinsey, *Private Contractors and the Reconstruction of Iraq: Transforming Military Logistics* (New York: Routledge, 2009), 66.

36. Caparini, "Domestic Regulation," 158–181.

37. Fainaru, "Where Military Rules."

38. "Operational Contract Support," Department of Defense, December 20, 2011, https://www.esd.whs.mil/Portals/54/Documents/DD/issuances/dodi/302041p.pdf?ver=2019-02-25-133949-097, 21–22. In 2007, DoD also established Armed Contractor Oversight Division to improve oversight of PMSCs by reducing incidents that might undermine the US mission in Iraq and to develop mechanisms that would hold PMSCs accountable for their conduct. See "Rebuilding Iraq: DOD and State Department Have Improved Oversight and Coordination of Private Security Contractors in Iraq, but Further Actions Are Needed to Sustain Improvements. Report to Congressional Committees," U.S. Government Accountability Office, July 2008, https://www.gao.gov/new.items/d08966.pdf, 10.

39. Fainaru, "Where Military Rules."

40. David Isenberg, "Private Military Contractors and U.S. Grand Strategy," International Peace Research Institute (PRIO) Report, 2009, https://www.files.ethz.ch/isn/109297/Isenberg%20Private%20Military%20Contractors%20PRIO%20Report%201-2009.pdf, 33. Although the State Department has established Tactical Operations Center in Iraq and improved oversight of contractors, there were shortages of agents in field. The State Department requested funding for additional Diplomatic Security agents to address this. See U.S. Government Accountability Office, "Rebuilding Iraq," 4.

41. Caparini, "Domestic Regulation."

42. Ibid.

43. Dickinson, "Contract as a Tool," 219.

44. Ibid.

45. Office of Inspector General, "Performance Evaluation of PAE Operations and Maintenance Support for the Bureau of International Narcotics and Law Enforcement Affairs' Counternarcotics Compounds in Afghanistan," February 2011, https://books.google.com/books?id=emQ4YKGANSYC&pg=PA7&lpg=PA7&dq=Performance+Evaluation+of+PAE+Operations+2011&source=bl&ots=eYJiItl4NH&sig=ACfU3U2Hb4PaWjV4NkaPDU35KWuSLAYTcQ&hl=en&ppis=_c&sa=X&ved=2ahUKEwjn8aHt4N7mAhVvRN8KHb0RDrkQ6AEwAXoECAoQAQ#v=onepage&q=Performance%20Evaluation%20of%20PAE%20Operations%202011&f=false, 3–7.

46. Benjamin Tkach, "Corporate Security and Conflict Outcomes" (PhD diss., Texas A&M University, 2013), 37–38.

47. Dickinson, "Contract as a Tool," 219–239.

48. Tkach, "Private Military and Security Companies, Contract Structure," 2.

49. Former Blackwater Employee, "Private Security."

50. Gary Schaub Jr. and Franke Volker, "Contractors as Military Professionals?," *Parameters: U.S. Army War College* 39, no. 4 (2009): 88–104.

51. Former Blackwater Employee, "Private Security."

52. Ibid.

53. International Code of Conduct Association, "International Code," 1–15.

54. "Procedures Article 12: Reporting, Monitoring and Assessing Performance and Compliance," The International Code of Conduct Association, accessed May 12, 2016, https://www.icoca.ch/sites/default/files/uploads/ICoCA-Procedures-Article-12-Monitoring.pdf, 4.

55. See, for example, "ISOA Code of Conduct," International Stability Operations Association, April 1, 2001, https://stability-operations.org/page/Code; Berenike Prem, *Private Military and Security Companies as Legitimate Governors: From Barricades to Boardrooms* (New York: Routledge, 2020), 131–151.

56. Matt Armstrong, "Private Security Company Association of Iraq," MountainRunner. us Blog, March 29, 2006, https://mountainrunner.us/2006/03/private_security_company_assoc/.

57. Nicole Jägers, "Will Transnational Private Regulation Close the Governance Gap?," in *Human Rights Obligations of Business: Beyond the Corporate Responsibility to Protect?* ed. Surya Deva and David Bilchitz (New York: Cambridge University Press, 2013), 295–329.

58. Jeffrey Herbst, "The Regulation of Private Security Forces," in *The Privatization of Security in Africa*, ed. Greg Mills and John Stremlau (Johannesburg: SAIIA Press, 1999), 107–129.

59. Dunigan, *Victory for Hire*, 29–35.

60. Virginia Newell and Benedict Sheehy, "Corporate Militaries and States: Actors, Interactions, and Reactions," *Texas International Law Journal* 41, no. 1 (2006): 71.

61. Richard Mulgan, "Comparing Accountability in the Public and Private Sectors," *Australian Journal of Public Administration* 59, no. 1 (2000): 87.

62. Anthony Vinci, *Armed Groups and the Balance of Power: The International Relations of Terrorists, Warlords, and Insurgents* (New York: Routledge, 2009): 51–52.

63. Thomas Gibbons-Neff and Jawad Sukhnayar, "The Taliban Have Gone High Tech. That Poses a Dilemma for the U.S.," *New York Times*, April 1, 2018, https://www.nytimes.com/2018/04/01/world/asia/taliban-night-vision.html..

64. Joseph Felter and Jake Shapiro, "Limiting Civilian Casualties as Part of a Winning Strategy: The Case of Courageous Restraint," *Daedalus* 146, no. 1 (2017): 48–50.

65. D. Scott Bennett, "Governments, Civilians, and the Evolution of Insurgency: Modeling the Early Dynamics of Insurgencies," *Journal of Artificial Societies and Simulations* 11, no. 4 (2008), http://jasss.soc.surrey.ac.uk/11/4/7.html, accessed July 15, 2019.

66. Steve Fainaru, "Iraq Contracts Face Growing Parallel War: As Security Work Increases So Do the Casualties," *Washington Post*, June 16, 2007, https://www.washingtonpost.com/wp-dyn/content/article/2007/06/15/AR2007061502602_3.html.

67. We argue that the shift in the balance of power can occur either in individual or co-deployed missions involving governmental forces together with PMSCs, though it is likely that co-deployed missions may offer more sustainable gains for the government, especially in cases where PMSCs are hired for a short-term contract. In cases where PMSCs' contracts terminate before the war ends, co-deployed missions offer more opportunities for the army to learn from private contractors in the field. This, in turn, may better prepare the army to conduct more-effective solo missions once the contractors depart and signal greater state capacity, an important element in stabilizing post-conflict environments. Nevertheless, the shift is likely to occur in either type of engagements. It is important to note the potential challenge of co-deployed missions that could delay the benefits of PMSCs' interventions in shifting the balance of power in favor of the government regardless of PMSCs' commitment to humanitarianism. Coordination and communication shortcomings have hindered US forces' responsiveness to the enemy in Iraq; issues of jealousy over higher pay and resentment over subordination to foreign contractors could also impact the effectiveness of the mission. See, for example, Dunigan, *Victory for Hire*, 78–79.

68. Tkach, "Private Military and Security Companies, Contract Structure," 2.

69. Allan R. Millett, Williamson Murray, and Kenneth H. Watman, "The Effectiveness of Military Organization," *International Security* 11, no. 1 (1986): 39–40.

70. Risa A. Brooks, "The Impact of Culture, Society, Institutions, and International Forces on Military Effectiveness," in *Creating Military Power: The Sources of Military Effectiveness*, ed. Risa Brooks and Elizabeth Stanley (Stanford, CA: Stanford University Press, 2007), 10.

71. Dunigan, *Victory for Hire*, 32.

72. Fitzsimmons, *Mercenaries*, 1–325.

73. Brooks, "The Impact of Culture," 10.

74. Dunigan, *Victory for Hire*, 35–41.

75. Brooks, "The Impact of Culture," 9–10.

76. Fitzsimmons, *Mercenaries*, 20–21.

77. Ibid.

78. Ibid.

79. Brooks, "The Impact of Culture," 13.

80. Dunigan, *Victory for Hire*, 74–76.

81. Ulrich Petersohn, "The Effectiveness of Contracted Coalitions: Private Security Contractors in Iraq," *Armed Forces & Society* 39, no. 3 (2013): 467.

82. Fitzsimmons, *Mercenaries*, 273–292.

83. Dunigan, *Victory for Hire*, 67.

84. Fitzsimmons, *Mercenaries*, 293–298, has an opposite finding. He shows that quantity and quality of military equipment was not decisive in PMSCs' ability to gain advantage in asymmetric conflicts.

85. Federation of European Accountants, "Integrity in Professional Ethics," September 2009, https://www.icaew.com/-/media/corporate/files/technical/ethics/integrity-in-professional-ethics-fee-discussion-paper.ashx?la=en.

86. Emile Ouedraogo, "Advancing Military Professionalism in Africa," African Center for Strategic Studies, July 2014, https://africacenter.org/wp-content/uploads/2016/06/ARP06EN-Advancing-Military-Professionalism-in-Africa.pdf, 4–5.

87. See, for example, United Nations Office of Human Resources Management, "Working Together, Putting Ethics into Work," November 20, 2008, https://docplayer.net/21479566-Working-together-putting-ethics-to-work-contents.html, 3.

88. Aaron Wildavsky and Naomi Caiden, *The New Politics of the Budgetary Process* (New York: Pearson/Longman, 2004), 173–174.

89. Aaron W. Dalenga and Brendan W. McGarry, "Defense Primer: DOD Transfer and Reprogramming Authorities," Congressional Research Service, June 7, 2019, https://fas.org/sgp/crs/natsec/IF11243.pdf.

90. There is a possibility that saved resources from PMSC overbilling could not be used for anything, therefore not contributing to military effectiveness. This, however, is unlikely. Although the law requires that unused resources in a given year are returned to the U.S. Treasury, DoD rarely does so. Instead the DoD sometimes shifts "one-year money" that it was intended to spend in one year to a pool of five-year money. After that, evidence shows that the money track is lost. See, for example, Dave Lindorff, "Exclusive: The Pentagon's Massive Fraud Exposed," *The Nation*, January 7, 2019, https://www.thenation.com/article/pentagon-audit-budget-fraud/. On transfer of funds to pay for the US-Mexico border wall, see for example, Andrew Taylor, "Trump's Use of Military Money for Border Wall Survives Senate Test," *Military Times*, October 18, 2019, https://www.militarytimes.com/news/pentagon-congress/2019/10/18/trumps-use-of-military-money-for-border-wall-survives-senate-test/.

91. Sandeep Saxena and Sami Ylaoutinen, "Managing Budgetary Viraments," International Monetary Fund, April 18, 2016, https://www.imf.org/en/Publications/TNM/Issues/2016/12/31/Managing-Budgetary-Virements-43850, 4–14.

92. Moshe Schwartz and Jennifer Church, "Department of Defense's Use of Contractors to Support Military Operations: Background, Analysis, and Issues for Congress," Congressional Research Service, May 17, 2013, https://fas.org/sgp/crs/natsec/R43074.pdf, 8.

93. Peter Singer, "The Dark Truth about Blackwater," Brookings, October 2, 2007, https://www.brookings.edu/articles/the-dark-truth-about-blackwater/.

94. David Isenberg, "PMC = Private Military Costs," *HuffPost*, January 28, 2013, http://www.huffingtonpost.com/david-isenberg/pmc-private-military-cost_b_2208825.html.

95. Daniel R. Blair and Evelyn R. Klemstine, "Afghan National Police Training Program: Lessons Learned during the Transition of Contract Administration," A Joint Audit by the Inspectors General of Department of State and Department of Defense, August 15, 2011, https://www.hsdl.org/?abstract&did=685111, 15.

96. Fitzsimmons, *Mercenaries*, 280–283.

97. International Committee of the Red Cross, "The Geneva Conventions of August 12, 1949," August 12, 1949, https://ihl-databases.icrc.org/ihl/INTRO/380, 170–171, 175, 179–180, 187.

98. Jason Lyall and Isaiah Wilson III, "Rage against the Machines: Explaining Outcomes in Counterinsurgency Wars," *International Organization* 63, no. 1 (2009): 70.

99. See, for example, Dave Kilcullen, "Two Schools of Classic Counterinsurgency," *Small Wars Journal* Blog, January 27, 2007, http://smallwarsjournal.com/blog/two-schools-of-classical-counterinsurgency; Bernard Finel, "An Alternative to COIN," *Armed Forces Journal*, February 1, 2010, http://armedforcesjournal.com/an-alternative-to-coin/.

100. Benjamin Valentino, Paul Huth, and Dylan Balch-Lindsay, "Draining the Sea: Mass Killings and Guerrilla Warfare," *International Organization* 58, no. 2 (2004): 385.

101. Finel, "An Alternative."

102. Kilcullen, "Two Schools."

103. Nathan Springer, "Stabilizing the Debate between Population-Centric and Enemy-Centric Counterinsurgency: Success Demands a Balanced Approach" (MA thesis, U.S. Army Command and General Staff College, 2011), 1–2.

104. Elisabeth Jean Wood, *Insurgent Collective Action and Civil War in El Salvador* (New York: Cambridge, 2003), 5–7.

105. Ibid.

106. Stathis Kalyvas and Matthew Kocher, "How 'Free' Is Freeriding in Civil Wars?: Violence, Insurgency, and the Collective Action Problem," *World Politics* 59, no. 2 (2007):186.

107. Singer, "The Dark Truth."

108. Lyall and Wilson, "Rage against the Machines," 73–75.

109. Regan, "Third-Party Interventions," 61.

110. Shanna Kirschner, "Families and Foes: Ethnic Civil War Duration" (PhD diss., University of Michigan, 2009), 29–30.

111. Barbara Walter, "The Critical Barrier to Civil War Settlement," *International Organization* 51, no. 3 (1997): 338–340.

112. Kirschner, "Families and Foes," 29–30.

113. Wood, *Insurgent Collective Action*, 2, 5, 7, 9–10.

114. Ibid., 18.

115. T. David Mason and Dale A. Krane, "The Political Economy of Death Squads: Toward a Theory of the Impact of State-Sanctioned Terror," *International Studies Quarterly* 33, no. 2 (1989): 188.

116. Ibid., 191.

117. Christopher Paul, Colin P. Clarke, and Beth Grill, *Victory Has a Thousand Words: Sources of Success in Counterinsurgency* (Santa Monica, CA: RAND, 2010), 97–98.

118. Lyall and Wilson, "Rage against the Machines," 67–106.

119. Ibid., 72–80.

120. Paul, Clarke, and Grill, *Victory*, 97–98

121. Ibid., 52.

122. Valentino, Huth, and Balch-Lindsay, "Draining the Sea," 402.

123. Yuri Zhukov, "A Theory of Indiscriminate Violence" (PhD diss., Harvard University, 2014), 30–38.

124. Lyall and Wilson, "Rage against the Machines," 72–78.

125. Valentino, Huth, and Balch-Lindsay, "Draining the Sea," 386.

126. Felter and Shapiro, "Limiting Civilian Casualties," 49.

127. Christopher Millson, "Comparing Counterinsurgency Tactics in Iraq and Vietnam," *Inquiries Journal* 3, no. 5 (2011): 1.

128. John Campbell, "Boko Haram Conflict Enters Counterinsurgency Phase as Nigeria Erects 'Fortresses,'" Council on Foreign Relations, December 7, 2017, https://www.cfr.org/blog/boko-haram-conflict-enters-counterinsurgency-phase-nigeria-erects-fortresses.

129. Noah Blaser, "Why Turkey's PKK Conflict Looms Larger than Ever in Local Election Aftermath," Foreign Policy Research Institute, April 9, 2019, https://www.fpri.org/article/2019/04/why-turkeys-pkk-conflict-looms-larger-than-ever-in-local-election-aftermath/.

130. The most notable exception to this today is the Syrian civil war, a conflict in which the government has systematically targeted civilians through indiscriminate attacks on the rebels. See, for example, Thanassis Cambanis, "The Logic of Assad's Brutality," *The Atlantic*, April 8, 2018, https://www.theatlantic.com/international/archive/2018/04/syria-chemical-weapons-assad-trump/557483/.

131. David Kilcullen, "Twenty-Eight Articles: Fundamentals of Company-Level Counterinsurgency," IO Sphere Joint Information Operations Center, Summer 2006, https://www.pegc.us/archive/Journals/iosphere_summer06_kilcullen.pdf, 29.

132. Springer, "Stabilizing the Debate," 118–119.

133. Sarah K. Cotton et al., *Hired Guns: Views about Armed Contractors in Operation Iraqi Freedom* (Santa Monica, CA: RAND, 2010), xiv.

134. Petersohn, "The Effectiveness," 478–479.

135. Daphna Canetti and Miriam Lindner, "Exposure to Political Violence and Political Behavior," in *Psychology of Change: Life Contexts, Experiences, and Identities*, ed. Katherine Reynolds and Nyla Branscombe (New York: Psychology Press, 2014), 77–94.

136. Jason Lyall, "Civilian Casualties, Radicalization, and the Effects of Humanitarian Assistance in Wartime Settings," Unpublished manuscript, April 5, 2015, http://aiddata.org/sites/default/files/lyall_2015_humanitarian_aid_afghanistan.pdf, 3.

137. Hal Brands, "Reform, Democratization, and Counter-Insurgency: Evaluating the US Experience in Cold War–Era Latin America," *Small Wars & Insurgencies* 22, no. 2 (2011): 308–309.

138. Dunigan, *Victory for Hire*, 71.

139. Cambanis, "The Logic of Assad's Brutality."

140. Avant, *The Market for Force*, 61.

141. Eugene Smith, "The New Condottieri and US Policy: The Privatization of Conflict and Its Implications," *Parameters* 32, no. 4 (2002): 104–119.

142. Dickinson, "Contract as a Tool," 221.

143. del Prado, "The Privatization of War"; Singer, *Corporate Warriors*, 45–47.

144. For differences in risk environments, see Ulrich Petersohn and Molly Dunigan, "Introduction," in *The Market for Force: Privatization of Security across World Regions*, ed. Molly Dunigan and Ulrich Petersohn (Philadelphia: University of Pennsylvania Press, 2015), 8. For insights about the Chinese PMSC market, see Jennifer Catallo, "China's Managed Market for Force," in *The Market for Force: Privatization of Security across World Regions*, ed. Molly Dunigan and Ulrich Petersohn (Philadelphia: University of Pennsylvania Press, 2015), 118–132.

145. Malyar Sadeq Azad, "Top Leaders Tied to Security Companies," Feral-Jundi, August 21, 2010, https://feraljundi.com/afghanistan-the-pscs-connected-to-karzais-family-and-close-associates/.

146. "Protocol Additional to the Geneva Conventions of 12 August 1949, and Relating to the Protection of Victims of International Armed Conflicts (Protocol 1)," International Committee of the Red Cross, June 8, 1977, https://ihl-databases.icrc.org/ihl/WebART/470-750057.

147. Today's PMSCs somewhat resemble more-organized private bodies that emerged in the late Middle Ages; see McFate, *The Modern Mercenary*, 27. The condottieri system consisted of permanent companies of armed militaries providing services to the highest bidder and was unlike the mercenaries of the 1970s and other ad hoc groups that provide services today. See Hyder Gulam, "The Rise and Rise of Private Military Companies" (Master's Thesis, Peace Operations Training Institute, 2005), 7–8.

148. Gulam, "The Rise," 7.

149. Singer, "Corporate Warriors: The Rise of the Privatized Military Industry and Its Ramifications," 191.

150. O'Brien, "What Should," 39.

151. Akcinaroglu and Radziszewski, "Private Military Companies," 795–821.

152. Our search terms included: the name of the country experiencing civil war and PMC/PMSC (e.g., Sierra Leone civil war and PMC/PMSC or Sierra Leone civil war and private military contractors/companies) and/or specific PMC/PMCS' names and civil wars (for example, Blackwater and civil war, Blackwater and Iraq, etc.).

153. Geoff Spencer, "Government Suspends Mercenary Contract after Violent Demonstration," Associated Press, March 20, 1997, https://apnews.com/2789a5b9f41da6faf45b531f607883e8.

154. Shantanu Chakrabarti, "Privatization of Security in the Post–Cold War Period," Institute for Defense Studies and Analyses, December 2009, https://idsa.in/system/files/Monograph_No2.pdf, 50.

155. Jean Houbert, "The Mascareignes, the Seychelles and the Chagos, Islands with a French Connection: Security in a Decolonized Indian Ocean," in *Political Economy of Small Tropical Islands: The Importance of Being Small*, ed. Helen M. Hintjens and Marilyn D. D. Newitt (Exeter, UK: University of Exeter Press, 1992), 87.

156. Meredith Reid Sarkees and Frank Wayman, eds., *Resort to War: 1816–2007* (Washington, DC: CQ Press, 2010).

157. Gleditsch et al., "Armed Conflict 1946–2001;" Lotta Themnér and Peter Wallensteen, "Armed Conflict, 1946–2012," *Journal of Peace Research* 50, no. 4 (2013): 509–521.

158. "UCDP/PRIO Armed Conflict Dataset Codebook Version 4–2009," International Peace Research Institute (PRIO), https://www.prio.org/Global/upload/CSCW/Data/UCDP/2009/Codebook_UCDP_PRIO%20Armed%20Conflict%20Dataset%20v4_2009.pdf, accessed May 10, 2019, 7.

159. While COW and UCDP/PRIO data sets are most commonly used in quantitative studies of civil wars, it is important to note their limitations. These include ambiguity in coding the start and end of wars precisely and the categorization of civil wars as distinct from other violent events such as coups, genocides, or even terrorism. For example, the COW Project defines war as an armed conflict that requires the participation of the government, effective resistance, and a thousand battle deaths. For definition, see Meredith Reid Sarkees, "The COW Typology of War: Defining and Categorizing Wars," The Correlates of War Project, https://correlatesofwar.org/data-sets/COW-war/the-cow-typology-of-war-defining-and-categorizing-wars/view, accessed on May 10, 2019, 1. Yet it is often difficult to guarantee a precise count of battle deaths amid unreliable or limited information. And the question always remains why 1,000 battle deaths and not some other number. Similar questions can be asked in reference to the battle death requirement found in the UCDP/PRIO data set. Furthermore, in his critique of the COW data set, Sambanis (2004) points to the difficulty in distinguishing civil wars from state-sponsored violence, or politicide, the latter not having to meet the requirement of effective resistance. See Nicholas Sambanis, "What Is Civil War? Conceptual and Empirical Complexities of an Operational Definition," *Journal of Conflict Resolution* 48, no. 6 (2004): 87–88. Lastly, some of the categorizations in the COW data set are debatable. The Ogaden war (1976–83), for example, is listed as an extra-systemic war in COW because the adversary is not a member of the interstate system, yet it shows all the other characteristics of a civil war. See Paul Collier and Anke Hoeffler, "Data Issues in the Study of Conflict," Conference Paper, June 6, 2001,https://pdfs.semanticscholar.org/e777/8066cc0569d63bac2548427b7630e6e2255b.pdf, 4. Overall, scholars agree that coding rules used to measure civil wars from battle deaths to onset and termination are rather ad hoc. When a war begins, and when it ends, has, however, implications for the phenomena we study such as duration. This discrepancy in coding rules results in considerable variation among the civil war data sets. While COW may code a war as terminated and thus having shorter durations due to its stringent criteria of what constitutes a civil war, UCDP/PRIO may count the same war as much longer based on a smaller battle death threshold of 25. Relying on both COW and UCDP/PRIO data sets is thus one way to test the robustness of our results on different specifications of onset, termination, and battle death thresholds.

160. While we include these conflicts in our analysis, we have an incomplete list of PMSCs that have intervened in both contexts. As such, we only list a sample of those interventions in our table for PMSC interventions into minor conflicts. Nevertheless, we include Iraq and Afghanistan because based on the incomplete

data we can already verify the presence of PMSC competition and publicly traded PMSCs for every conflict year in our data set.

161. Avant, *The Market for Force*, 225–226.
162. Olasupo, "Nigeria: How Buhari Stopped Us from Fighting"; de Nevers, "Private Military Contractors and Changing Norms," 69–74.
163. Andreas Schedler, "Mexico: Transition to Civil War Democracy," in *Politics in the Developing World*, ed. Peter Burnell, Lise Rakner, and Vicky Randall (New York: Oxford University Press, 2014), 342.
164. See, for example, Stathis Kalyvas, "'New' and 'Old' Civil Wars: A Valid Distinction?," *World Politics* 54, no. 1 (2001): 103–104; Michael Ross, "A Closer Look at Oil, Diamonds, and Civil War," *Annual Review of Political Science* 9 (2006): 265.
165. Francisco Gutiérrez-Sanin, "Criminal Rebels? Discussion of Civil War and Criminality from the Colombian Experience," *Politics & Society* 32, no. 2 (2004): 258.e
166. Javier Osorio, "Democratization and Drug Violence in Mexico," Working Paper, https://eventos.itam.mx/sites/default/files/eventositammx/eventos/aadjuntos/2014/01/democratizacion_and_drug_violence_osorio_appendix_1.pdf, accessed on October 29, 2018, 4.
167. Sambanis, "What Is Civil War?," 855.
168. Paul Collier, Anke Hoeffler, and Måns Söderbom, "On the Duration of Civil War," *Journal of Peace Research* 41 no. 3 (2004): 257.
169. Sambanis, "What Is Civil War?," 831; David E. Cunningham, Kristian Skrede Gleditsch, and Idean Salehyan, "It Takes Two: A Dyadic Analysis of Civil War Duration and Outcome," *Journal of Conflict Resolution* 53, no. 2 (2009): 583.
170. See Chapter 5 for more in-depth discussion of our coding rules for the analysis.
171. Singer, *Corporate Warriors*, 88–100.
172. Avant, *The Market for Force*, 17.
173. Christopher Kinsey, *Corporate Soldiers and International Security: The Rise of Private Military Companies* (London: Routledge, 2006), 10.
174. Dunigan, *Victory for Hire*, 13.
175. See, for example, Regan, "Third Party Interventions," 55–73; Patrick M. Regan and Aysegul Aydin, "Diplomacy and Other Forms of Intervention in Civil Wars," *Journal of Conflict Resolution* 50, no. 5 (2006): 736–756; Jacob D. Kathman, "Civil War Diffusion and Regional Motivations for Intervention," *Journal of Conflict Resolution* 55, no. 6 (2011): 847–876; and Noel Anderson, "Competitive Intervention, Protracted Conflict, and the Global Prevalence of Civil War," *International Studies Quarterly* 63, no. 3 (2019): 692–706.

# Chapter 3

1. Musah and Fayemi, "Africa in Search of Security," 22, 23, 27; Isenberg, "Private Military Contractors," 11.
2. James M. O'Brien III, "Private Military Companies: An Assessment" (Master's thesis, Naval Postgraduate School, 2008), 39.

3. Cockayne, "Make or Buy?," 196–217.

4. Jaroslav Tir and Michael Jasinski, "Domestic-Level Diversionary Theory of War: Targeting Ethnic Minorities," *Journal of Conflict Resolution* 52, no. 5 (2008): 1–2.

5. Alexander Griffing, "How Assad Helped Create ISIS to Win in Syria and Got Away with the Crime of the Century," *Haaretz*, October 7, 2018, https://www.haaretz.com/middle-east-news/syria/MAGAZINE-iran-russia-and-isis-how-assad-won-in-syria-1.6462751.

6. Tir and Jasinski, "Domestic-Level Diversionary Theory," 643–644.

7. "German Arms Firm Ends Blackwater Deal after TV Report," *Deutsche Welle*, February 19, 2008, https://www.dw.com/en/german-arms-firm-ends-blackwater-deal-after-tv-report/a-3135177.

8. O'Brien, "Private Military Companies and African Security," 44.

9. Sebastian Drutschmann, "Motivation, Markets and Client Relations in the British Private Security Industry" (PhD diss., King's College, 2014), 211.

10. Fitzsimmons, *Private Security Companies during the Iraq War*, 213.

11. Allison, "Informal but Diverse," 92.

12. Akcinaroglu and Radziszewski, "Private Military Companies," 795–821.

13. Jeremy Scahill, *Blackwater: The Rise of the World's Most Powerful Mercenary Army* (New York: Nation Books, 2007), 46.

14. For a specific list of all international PMSCs that have intervened in Iraq at any point from 2003–2019, see Appendix 4.1.

15. One might argue that an increase in the number of PMSCs operating in a given conflict, and thus the presence of local competition, is offset by an increase in the buyers of PMSC services. This is rarely the case, however. Our data show that in most instances there is a single buyer, the government, in a given conflict. The most notable exceptions are the civil war in Peru, recent wars in Iraq and Afghanistan, and the conflict in Colombia where both the US government and Iraqi/Colombian/Peruvian/Afghan governments hired PMSCs.

16. Former Pacific Architects and Engineers Employee, "Competition and Effectiveness," interview by the author, Elizabeth Radziszewski, April 29, 2012.

17. Dunigan, *Victory for Hire*, 73.

18. Carmola, "PMSCs and Risk in Counterinsurgency," 135.

19. Paul, Clarke, and Grill, *Victory*, 97–98.

20. Carmola, "PMSCs and Risk in Counterinsurgency," 149.

21. Claire Vallings and Magui Moreno-Torres, "Drivers of Fragility: What Makes States Fragile?," Department for International Development UK, April 2005, https://webcache.googleusercontent.com/search?q=cache:H6nEcSvaeX0J:https://ageconsearch.umn.edu/bitstream/12824/1/pr050007.pdf+&cd=15&hl=en&ct=clnk&gl=us, 6.

22. Avant, *The Market for Force*, 43–45.

23. Adam Nossiter, "After Vote in Congo, Talk of Resistance," *New York Times*, December 7, 2011, https://www.nytimes.com/2011/12/08/world/africa/after-vote-in-congo-talk-of-resistance.html?mtrref=www.google.com&gwh=C6E700DF52B47DD302CE73A39BB94DA5&gwt=pay&assetType=REGIWALL.

24. Michael Porter, "The Five Competitive Forces That Shape Strategy," *Harvard Business Review*, January 2008, https://hbr.org/2008/01/the-five-competitive-forces-that-shape-strategy. Other factors that affect competitive intensity include: threat of substitute products or services, bargaining power of suppliers and customers, and rivalry among existing competitors.

25. Sebastian Drutschmann, "Informal Regulation: An Economic Perspective on the Private Security Industry," in *Private Military and Security Companies: Chances, Problems, Pitfalls, and Prospects*, ed. Thomas Jager and Gerhard Kummel (Wiesbaden: VS Verlag für Sozialwissenschaften, 2007), 449–450.

26. Hal Varian, "Monitoring Agents with Other Agents," *Journal of Institutional and Theoretical Economics* 146, no. 1 (1990): 153.

27. Leticia Armendariz, "Human Rights Abuses and Monitoring," interview by the author, Seden Akcinaroglu, January 10, 2014.

28. Charles Glass, "The Warrior Class: A Golden Age for the Freelance Soldier," *Harper's Magazine*, April 2012, https://harpers.org/archive/2012/04/the-warrior-class/.

29. Anne Mette Christiansesn, Thomas Trier Hansen, and Riikka Poukka, "Nordic Human Rights Study," Deloitte, 2015, https://www2.deloitte.com/content/dam/Deloitte/no/Documents/risk/Nordic-Human-Rights-Study.pdf, 11.

30. Spear, *Market Forces*, 44.

31. Howe, "Private Security Forces," 316.

32. Petersohn, "The Social Structure of the Market," 273.

33. Drutschmann, "Informal Regulation," 443–455.

34. Although the focus of this book is on the impact of multiple companies on their behavior, other actors, including local communities and NGOs, also contribute to creating a wide range of informal mechanisms, collectively creating a stronger mechanism of accountability.

35. Drutschmann, "Informal Regulation," 447–449.

36. Ibid., 448.

37. Kevin C. Kennedy, "A Critical Appraisal of Criminal Deterrence Theory," *Dickinson Law Review* 88, no. 1 (1983–1984): 5.

38. Gary Kleck et al., "The Missing Link in General Deterrence Research," *Criminology* 43, no. 3 (2005): 626.

39. Adam Zagorin, "A 'Mutiny' in Kabul: Guards Allege Security Problems Have Put Embassy at Risk," Project on Government Oversight, January 17, 2013, https://www.pogo.org/investigation/2013/01/mutiny-in-kabul-guards-allege-security-problems-have-put-embassy-at-risk/.

40. Apuzzo, Matt, "Ex Blackwater Guards Given Long Terms after Killing Iraqis," *New York Times*, April 14, 2015, http://www.nytimes.com/2015/04/14/us/ex-blackwater-guards-sentenced-to-prison-in-2007-killings-of-iraqi-civilians.html.

41. Former Blackwater Employee, "Private Security."

42. Doug Brooks, "Accountability and Private Contractors," interview conducted by the author, Elizabeth Radziszewski, April 2016.

43. Former PMSC employee, "Accountability and Private Contractors," interview conducted by the author, Elizabeth Radziszewski, September 28, 2012.

44. Brooks, "Accountability."
45. Former Pacific Architects and Engineers Employee, "Competition."
46. McFate, *The Modern Mercenary*, 118.
47. Lars Lindblom, "Dissolving the Moral Dilemma of Whistleblowing," *Journal of Business Ethics* 76, no. 4 (2007): 415.
48. Mark Somers and Jose C. Casal, "Type of Wrongdoing and Whistle-Blowing: Further Evidence That Type of Wrongdoing Affects the Whistle-Blowing Process," *Public Personnel Management* 40, no. 2 (2011): 160; Barbara Masser and Rupert Brown, "When Would You Do It? An Investigation into the Effects of Retaliation, Seriousness of Malpractice and Occupation on Willingness to Blow the Whistle," *Journal of Community and Applied Social Psychology* 6, no. 2 (1996): 129.
49. Cotton et al., *Hired Guns*, xiv.
50. Funmi Olonisakin, "Arresting the Tide of Mercenaries: Prospects for Regional Control," in *Mercenaries: An African Security Dilemma*, ed. Abdel-Fatau Musah and J. 'Kayode Fayemi (Sterling, VA: Pluto Press, 2000), 236.
51. Michael Scheimer, "Separating Private Military Companies from Illegal Mercenaries in International Law: Proposing an International Convention for Legitimate Military and Security Support That Reflects Customary International Law," *American University International Law Review* 24, no. 3 (2009): 636.
52. "Directory of Private Military Companies," PrivateMilitary.org, http://www.privatemilitary.org/private_military_companies.htm, accessed May 10, 2018.
53. Khareen Pech, "Executive Outcomes—A Corporate Conquest," in *Peace, Profit or Plunder? The Privatization of Security in War-Torn African Societies*, ed. Jakkie Cilliers and Douglas Fraser (Pretoria, South Africa: Institute for Security Studies, 1999), 84.
54. Alan Axelrod, *Mercenaries: A Guide to Private Armies and Private Military Companies* (Thousand Oaks, CA: CQ Press, 2014), 267–268.
55. Ibid.
56. David J. Francis, "Mercenary Intervention in Sierra Leone: Providing National Security or International Exploitation?," *Third World Quarterly* 20, no. 2 (1999): 322.
57. Avant, *The Market for Force*, 98.
58. Pech, "Executive Outcomes," 86.

# Chapter 4

1. Marc von Boemcken, *Between Security Markets and Protection Rackets: Formations of Political Order* (Opladen, Germany: Budrich UniPress Ltd., 2013), 161.
2. Porter, "The Five Competitive Forces."
3. Daniel Bergner, "The Other Army," *New York Times Magazine*, August 14, 2005, https://www.nytimes.com/2005/08/14/magazine/the-other-army.html.
4. Richard Wahman, "Iraq Security Firm Joins Bidding for Wall Street's Favorite Detective Agency," *The Guardian*, March 13, 2010, https://www.theguardian.com/business/2010/mar/14/kroll-control-risks-bidding-war.

5. Alex Klein, "U.S. Army Awards Iraq Security Work to British Firm," *Washington Post*, September 14, 2007, http://www.washingtonpost.com/wp-dyn/content/article/2007/09/13/AR2007091302237.html.

6. "An Uneasy Relationship: U.S. Reliance on Private Security Firms in Overseas Operations," Committee on Homeland Security and Governmental Affairs, US Senate, February 27, 2008, https://fas.org/irp/congress/2008_hr/psc.pdf, 14.

7. Ibid.

8. Laura Peterson and Phillip van Niekerk, "Privatizing Combat—The New World Order," The Public i-Center for Public Integrity, November/December 2002, http://www.thirdworldtraveler.com/War_Peace/Privatizing_Combat.html.

9. Charles J. Dunar III, Jared J. Mitchell, and Donald L. Robbins III, "Private Military Industry Analysis: Private and Public Companies" (MBA Thesis, Naval Postgraduate School, 2007), 39–40. Although acquired by another company, such PMSCs are still recognizable actors in the industry. Despite being owned by L-3 Communications (renamed L3 Technologies in 2016), MPRI is registered under its own name as MPRI, an L-3 Company. Other types of publicly traded PMSCs, a substantial minority, are new divisions from larger corporations not traditionally associated with security and defense or spin-offs, companies that broke away from a parent company. For example, Kellogg Brown and Root (KBR) started after breaking away from Halliburton in 1998. See Dunar, Mitchell, and Robbins, "Private Military," 39–40.

10. This does not mean that the media devote the same level of coverage to publicly traded PMSCs as they do to public forces, rather that publicly traded PMSCs are likely to generate more extensive coverage than privately held PMSCs. Partially as a response to greater public demand and partially to greater availability of information about the companies' financial disclosures, journalists are in a much better position to focus on publicly traded PMSCs than on privately held firms in the same industry.

11. Arlene Weintraub, "10 Questions to Ask before Taking Your Company Public," *Entrepreneur*, July 25, 2013, http://www.entrepreneur.com/article/227487.

12. Cosmina Lelia Voinea and Hans van Kranenburg, "Media Influence and Firms Behavior: A Stakeholder Management Perspective," *International Business Research* 10, no. 10 (2017): 1.

13. Les Brorsen, "Looking Behind the Declining Number of Public Companies," Harvard Law School Forum on Corporate Governance and Financial Regulation, May 18, 2017, https://corpgov.law.harvard.edu/2017/05/18/looking-behind-the-declining-number-of-public-companies/.

14. Ibid.

15. Clive Chajet, "Corporate Reputation as a Strategic Asset: Corporate Reputation and the Bottom Line," *Corporate Reputation Review* 1 (1997): 22.

16. Ibid.

17. Dallas Hanson and Helen Stuart, "Failing the Reputation Management Test: The Case of BHP, the Big Australian," *Corporate Reputation Review* 4, no. 2 (2001): 132.

18. David Isenberg, "Best of Luck to You Ms. Burke," CATO Institute, May 14, 2010, http://www.cato.org/publications/commentary/best-luck-you-ms-burke-0.

19. Scott Eden, "DynCorp Ends Public Experiment," *The Street*, April 13, 2010, https://www.thestreet.com/investing/stocks/dyncorp-ends-public-experiment-10723699.

20. Christopher Hinton, "DynCorp to Go Private in $1.5 Billion Deal with Cerberus," *Market Watch*, April 12, 2010, https://www.marketwatch.com/story/cerberus-to-take-dyncorp-private-for-15-billion-2010–04-12.

21. Eden, "DynCorp Ends."

22. Hinton, "DynCorp to Go Private."

23. Antony Barnett and Solomon Hughes, "British Firm Accused in UN 'Sex Scandal,'" *The Guardian*, July 28, 2001, https://www.theguardian.com/world/2001/jul/29/unitednations.

24. Kevin Haeberle and M. Todd Henderson, "Making a Market for Corporate Disclosure," *Yale Journal of Regulation* 35, no. 2 (2018): 427.

25. Ibid; Dain Donelson, Justin J. Hopkins, and Christopher G. Yust, "The Cost of Disclosure Regulation: Evidence from D & O Insurance and Non-Meritorious Securities Litigation," The Columbia Law School Blue Sky Blog, February 22, 2018, http://clsbluesky.law.columbia.edu/2018/02/22/the-cost-of-disclosure-regulation-evidence-from-do-insurance-and-non-meritorious-securities-litigation/.

26. Brorsen, "Looking Behind."

27. Heather Elms and Robert A. Phillips, "Private Security Companies and Institutional Legitimacy: Corporate and Stakeholder Responsibility," *Business Ethics Quarterly* 19, no. 3 (2009): 416.

28. Sorocha Macleod and Rebecca Dewinter-Schmitt, "Certifying Private Security Companies: Effectively Ensuring the Corporate Responsibility to Respect Human Rights?," *Business and Human Rights Journal* 4, no. 1 (2019): 55–60. On PMSCs' acceptance of ethical norms see Christopher Kinsey, "Private Security Companies and Corporate Social Responsibility," In *Private Military and Security Companies: Ethics, Policies and Civil-Military Relations*, ed. Andrew Alexandra, Deane-Peter Baker and Marina Caparini (New York, NY: Routledge, 2008), 70–86.

29. Former Pacific Architects and Engineers Employee, "Private Military Companies," interview by the author, Elizabeth Radziszewski, April 10, 2012.

30. Committee on Homeland Security and Governmental Affairs, "An Uneasy Relationship," 13.

31. Former Pacific Architects and Engineers Employee, "Private Military."

32. Committee on Homeland Security and Governmental Affairs, "An Uneasy Relationship," 13.

33. Steve Fainaru, "Iraq Contracts Face Growing Parallel War: As Security Work Increases So Do the Casualties," *Washington Post*, June 16, 2007. https://www.washingtonpost.com/wp-dyn/content/article/2007/06/15/AR2007061502602_3.html.

34. Avant, *The Market for Force*, 63–64.

35. Pech, "Executive Outcomes," 94.

36. James Kouzes and Barry Posner, *The Leadership Challenge* (San Francisco: Jossey Bass, 2002), 80–81.

37. Former Pacific Architects and Engineers Employee, "Private Military."

38. Ibid.

39. Ibid.

40. John B. Thompson, "The New Visibility," *Theory, Culture and Society* 22, no. 6 (2005): 31.

41. del Prado, "The Privatization of War"; Singer, *Corporate Warriors*, 45–47.

42. See Appendix 4.1 for a list of companies. Our list includes six private companies that were acquired by publicly traded PMSCs. Our decision to count those six companies as public rests on the premise that such companies would be held accountable to higher standards by the parent company itself whose reputation would be damaged should its subsidiary behave in unethical ways. Nevertheless, we also tested the robustness of our results by adding a control variable "acquisition by a public corporate company" for those six observations and our results remained the same.

43. Examples of search terms in connection with PMSCs: "fraud," "killed," "wounded," "accused," "alleged," "allegations," "scandal," "human rights violations," "crime," "illegal," "stolen," "bribed," "kickbacks," "sued," "corruption."

44. Jordi Palou-Loverdos and Leticia Armendariz, "The Privatization of Warfare, Violence and Private Military and Security Companies: A Factual and Legal Approach to Human Rights Abuses by PMSC in Iraq," International Institute for Non Violent Action, 2011, https://novact.org/wp-content/uploads/2012/09/The-privatization-of-warfare.pdf, 120–261.

45. We eliminated any duplicate news from this count.

46. Spencer Hsu, "Blackwater Guard's Retrial for Murder in 2007 Iraq Massacre Goes to U.S. Jury," *Washington Post*, August 8, 2018, https://www.washingtonpost.com/local/public-safety/blackwater-guards-retrial-for-murder-in-2007-iraq-massacre-goes-to-us-jury/2018/08/08/710671a8-98cd-11e8-843b-36e177f3081c_story.html.

47. Apuzzo, "Ex Blackwater Guards."

48. We used a negative binomial model, a model that is recommended when the data are overdispersed, in other words, when the conditional variance exceeds the conditional mean. Test results showed that the model is superior to other count models such as Poisson and zero inflated models.

49. Elke Krahmann, "NATO Contracting in Afghanistan: The Problem of Principal-Agent Networks," *International Affairs* 92, no. 6 (2016): 1401–1402.

50. See Appendix 4.2 for variable coding.

51. Robert Beckhusen, "A Decade Later, Contractor Pays Out Millions for Iraq Prisoner Abuse," *Wired*, January 9, 2013, https://www.wired.com/2013/01/torture-settlement/.

52. James Boxell, "Competition Hits Security Groups in Iraq," *Financial Times*, November 5, 2005.

53. Martin Smith, "Private Warriors," Frontline, 2005http://www.shoppbs.pbs.org/wgbh/pages/frontline//shows/warriors/etc/script.html, accessed July 18, 2019.

54. Dickinson, "Contract as a Tool," 219.

55. "Iraq Reconstruction: Lessons Learned in Contracting," Committee on Homeland Security and Governmental Affairs, US Senate, August 2006, https://www.govinfo.gov/content/pkg/CHRG-109shrg29761/html/CHRG-109shrg29761.htm, 3–5.

56. Mayer, "DoD Monitoring."

57. Tilman Rodenhauser and Jonathan Cuenoud, "Speaking Law to Business: 10-Year Anniversary of the Montreux Document on PMSCs," Law and Conflict, September 17, 2018, https://blogs.icrc.org/law-and-policy/2018/09/17/speaking-law-business-10-year-anniversary-montreux-document-pmscs/; Prem, Private Military and Security, 131–151.

58. Deborah Avant, "Pragmatic Networks and Transnational Governance of Private Military and Security Services," International Studies Quarterly 60, no. 2 (2016): 336.

59. We do not include all of the variables in one model. Our sample size is not large enough to offer reliable insights with too many control variables. As such, we keep several variables unchanged in all three models, while including and excluding a variety of different controls.

60. Singer, Corporate Warriors, 133.

61. Libby McCarthy, "New Report Reveals 86% of U.S. Consumers Expect Companies to Act on Social, Environmental Issues," Sustainable Brands, 2017, accessed on April 5, 2019, https://sustainablebrands.com/read/marketing-and-comms/new-report-reveals-86-of-us-consumers-expect-companies-to-act-on-social-environmental-issues.

62. Steve Fainaru, "Cutting Costs, Bending Rules, and a Trail of Broken Lives," Washington Post, July 29, 2007, https://www.washingtonpost.com/wp-dyn/content/article/2007/07/28/AR2007072801407.html.

63. Petersohn and Dunigan, "Introduction," 4–12.

64. Catallo, "China's Managed," 118.

65. Kristina Mani, "Diverse Markets for Force in Latin America: From Argentina to Guatemala," in The Market for Force: Privatization of Security Across World Regions, ed. Molly Dunigan and Ulrich Petersohn (Philadelphia: University of Pennsylvania Press, 2015), 29–34.

66. Elke Krahmann, "Choice, Voice and Exit: Consumer Power and the Self-Regulation of the Private Security Industry," European Journal of International Security 1, no. 1 (2016): 38.

67. Cusumano, "Regulating," 10.

68. Boxell, "Competition."

69. Richard Norton-Taylor and Richard Wray, "Boss Quits ArmorGroup after Iraq Problems," The Guardian, November 28, 2007, https://www.theguardian.com/business/2007/nov/28/iraq.

70. Petersohn, "The Effectiveness of Contracted Coalitions," 467.

71. Loverdos and Armendariz, "The Privatization of Warfare," 49.

72. Krahmann, "NATO Contracting," 1402; Sean McFate, "America's Addiction to Mercenaries," The Atlantic, August 12, 2016, https://www.theatlantic.com/international/archive/2016/08/iraq-afghanistan-contractor-pentagon-obama/495731/.

73. Former Pacific Architects and Engineers Employee, "Private Military."

74. McFate, The Modern Mercenary, 24–25.

75. Mayer, "Private Security."

76. Krahmann, "Choice, Voice," 39, 46.

77. McFate, The Modern Mercenary, 163.

78. Peter Beaumont, "Abu Ghraib Abuse Firms Are Rewarded," *The Guardian*, January 15, 2005, https://www.theguardian.com/world/2005/jan/16/usa.iraq; Sue Sturgis, "Blackwater's Iraq Contract Renewed despite Ongoing Massacre Probe," *Facing South*, April 7, 2008, https://www.facingsouth.org/2008/04/blackwaters-iraq-contract-renewed-despite-ongoing-massacre-probe.html.

79. Sue Pleming and Andy Sullivan, "FBI Takes Lead in Blackwater Investigation," Reuters, October 4, 2007, https://www.reuters.com/article/us-usa-iraq-blackwater/fbi-takes-lead-in-blackwater-investigation-idUSN0430861520071005; "Blackwater Banned from Iraq," Al Jazeera, January 30, 1999, https://www.aljazeera.com/news/middleeast/2009/01/2009129103918814445.html.

80. Mark Schleub, "Alan Grayson, Candidate and Judge, Gets Sanctioned from the Judge," *Orlando Sentinel*, October 30, 2008, http://articles.orlandosentinel.com/2008–10-30/news/grayson30_1_grayson-iraq-war-judge.

# Chapter 5

1. Elise Labott, " U.S. Will Not Renew Iraq Contract with Blackwater," CNN, January 30, 2009, http://edition.cnn.com/2009/WORLD/meast/01/30/us.blackwater.contract/.

2. "Security Contractor Boosts Ethics Practices," *The Business Ethics Blog*, July 29, 2009, https://businessethicsblog.com/2009/07/, 1.

3. This statistical model is not only preferable to OLS (Ordinary Least Squares Regression), which fails to address time dependence when analyzing the duration of war, but is also preferred to other parametric models including Weibull estimation. Unlike the Cox model, the Weibull distribution model applies restrictive assumptions on the distribution of the baseline hazard. This implies that using a Cox model with flexible baseline hazard is recommended for our analysis. The basic specification for Cox model is:

$$h_i(t) = h_0(t)\exp(\beta_1 x_{11} + \beta_2 x_{1k} + \cdots + \beta_k x_{1k}) \text{ or } h(t) = h_0(t)e^{x\beta}$$

where $h_0(t)$ is the baseline hazard, $\beta$'s are slope parameters and x's are independent variables. In this semi-parametric model, the baseline hazard, $h_0(t)$, remains unspecified while the covariates enter the model linearly. The hazard model is based on yearly transition rates from war to peace. We right-censor conflicts that are still ongoing by the end point of our data, which is December 31, 2008. That is, the survival of ongoing observations remains unknown. There are no censored observations in our data set when we use the COW Intra-state War data set, meaning all of the conflicts have ended by 2008, while we have 33 censored cases with the UCDP/PRIO Armed Conflict data set. In Cox models, the hazard ratios following each estimation are interpreted according to whether or not they exceed 1; those ratios that are greater than 1 imply that greater values of the variable increase the risk of failure, or in this case the termination of conflict. Higher values of the variables with hazard ratios less than 1 contribute to the continuation of war, that is, they lead to prolonged conflicts. The Cox model is based on an assumption of proportional hazards. This means that the impact of each variable on conflict duration remains the same across time. We performed diagnostics checks to test this assumption in our models. The proportionality assumption was violated for three control variables, *Polity, Mountainous Terrain,* and *Rebel Support,* for major and minor conflicts in some models. Thus, in our analysis we include these time-varying covariates when appropriate. Time-varying covariates mean that the variables' effect on the duration of war will change with the passage of time.

4. See Appendix 5.1 for more information about control variables and coding.
5. Barbara Walter, "Bargaining Failures and Civil War," *Annual Review of Political Science* 12 (2009): 253.
6. Phillippe LeBillon, "The Political Ecology of War: Natural Resources and Armed Conflicts," *Political Geography* 20, no. 5 (2001): 570; Richard Auty, "Natural Resources and Civil Strife: A Two-Stage Process," *Geopolitics* 9, no. 1 (2004): 42–43.
7. Clayton Thyne, "Third Party Intervention and the Duration of Civil Wars: The Role of Unobserved Factors," Working Paper, 2008, https://www.researchgate.net/publication/228429944_Third_Party_Intervention_and_the_Duration_of_Civil_Wars_The_Role_of_Unobserved_Factors, 5, accessed January 12, 2012.
8. Seden Akcinaroglu, "Rebel Interdependencies and Civil War Outcomes," *Journal of Conflict Resolution* 56, no. 5 (2012): 885; David E. Cunningham, "Veto Players and Civil War Duration," *American Journal of Political Science* 50, no. 4 (2006): 887.
9. Philip Hultquist, "Power Parity and Peace? The Role of Relative Power in Civil War Settlement," *Journal of Peace Research* 50, no. 5 (2013): 626.
10. Michael L. Ross, "What Do We Know about Natural Resources and Civil War?" *Journal of Peace Research* 41, no. 3 (2004): 350; Collier, Hoeffler, and Söderbom, "On the Duration," 268.
11. Virginia Page Fortna, "Do Terrorists Win? Rebels' Use of Terrorism and Civil War Outcomes," *International Organization* 69, no. 3 (2015): 524, 550.
12. See Appendix 5.1 for discussion of the selection bias test.
13. Cunningham, "Veto Players," 871.
14. Collier, Hoeffler, and Söderbom, "On the Duration," 262.
15. In the conflicts of Angola, Sierra Leone, and Democratic Republic of the Congo, Executive Outcomes (EO) was hired to help the government, along with companies such as Sandline, Saracen, Bridge International, Teleservices, Lifeguard, Ibis Air International, Cape International, Panasec, Stabilico, and Omega Support. These companies were considered to be subsidiaries of EO, affiliated companies with EO, or PMSCs that were formed by ex-managers of EO. To avoid overestimating the competitive environment in these conflicts, we ran the analysis excluding those three conflicts. Our results remained the same.
16. Regan, "Third-Party Interventions," 72.
17. Cunningham, Skrede Gleditsch, and Salehyan, "It Takes Two," 574–575.
18. Regan, "Third-Party Interventions," 55; Balch-Lindsay and Enterline, "Killing Time," 615; Akcinaroglu and Radziszewski, "Expectations," 349.
19. Cunningham, Skrede Gleditsch, and Salehyan, "It Takes Two," 579–581.
20. James D. Fearon, "Why Do Some Civil Wars Last So Much Longer Than Others?" *Journal of Peace Research* 41, no. 3 (May 2004): 298.

21. Jeffrey Herbst, "African Militaries and Rebellion: The Political Economy of Threat and Combat Effectiveness," *Journal of Peace Research* 41, no. 3 (2004): 357.

22. Alan Heston, Robert Summers, and Bettina Aten, "Penn World Table Version 6.3," Center for International Comparison of Production Income, and Prices, University of Pennsylvania, August 2009, http://datacentre2.chass.utoronto.ca/pwt/, accessed June 8, 2013.

23. James D. Fearon and David D. Laitin, "Ethnicity, Insurgency, and Civil War," *American Political Science Review* 97, no. 1 (2003): 88.

24. Paul Collier and Anke Hoeffler, "Greed and Grievance in Civil War," *Oxford Economic Papers* 56, no.4 (2004): 563.

25. "Polity IV Project: Political Regime Characteristics and Transitions, 1800–2013," Center for Systemic Peace, last modified June 6, 2014, https://www.systemicpeace.org/polity/polity4.htm.

26. Balch-Lindsay and Enterline, "Killing Time," 622.

27. Cunningham, Skrede Gleditsch, and Salehyan, "It Takes Two," 579.

28. Chaim Kaufmann, "Possible and Impossible Solutions to Ethnic Civil Wars," *International Security* 20, no. 4 (1996): 137; Cunningham, Skrede Gleditsch, and Salehyan, "It Takes Two," 584.

29. Kaufmann, "Possible and Impossible," 137.

30. Bethany Lacina and Nils Petter Gleditsch, "Monitoring Trends in Global Combat: A New Dataset of Battle Deaths," *European Journal of Population* 21, no. 2–3 (2005): 145–166.

31. T. David Mason, Joseph P. Weingarten, and Patrick J. Fett, "Win, Lose, or Draw: Predicting the Outcome of Civil Wars," *Political Research Quarterly* 52, no. 2 (1999): 239.

32. Balch-Lindsay and Enterline, "Killing Time," 624.

33. Fearon and Laitin, "Ethnicity, Insurgency, and Civil War," 81.

34. Halvard Buhaug and Scott Gates, "The Geography of Civil War," *Journal of Peace Research* 39, no. 4 (2002): 417–433.

35. Cunningham, Skrede Gleditsch, and Salehyan, "It Takes Two," 579.

36. Balch-Lindsay and Enterline, "Killing Time," 624, 636.

37. Regan, "Third-Party Interventions," 63, 69.

38. World Bank, "World Development Indicators," http://databank.worldbank.org/ddp/home.do?Step=12&id=4&CNO=2, accessed on October 10, 2019.

39. Sambanis, "What Is Civil War?" 822.

40. Macartan Humphreys, "Natural Resources, Conflict, and Conflict Resolution: Uncovering the Mechanisms," *Journal of Conflict Resolution* 49, no. 4 (2005): 522–523.

41. Walter Enders, Todd Sandler, and Khusrav Gaibulloev, "Domestic versus Transnational Terrorism: Data, Decomposition, and Dynamics," *Journal of Peace Research* 48, no. 3 (2011): 319–337.

42. Regan, "Third-Party Interventions," 61.

43. In both cases, we predicted concordance probability without the time-varying covariates.

# Chapter 6

1. Tkach, "Private Military and Security Companies, Contract Structure," 291–311.

2. Petersohn, "Private Military and Security Companies," 1046–1072.

3. Dunigan, *Victory for Hire*, 156.

4. Fitzsimmons, *Mercenaries*, 47–273.

5. Petersohn, "Private Military and Security Companies," 1046–1072.

6. Petersohn, "The Impact of Mercenaries," 191–215.

7. Akcinaroglu and Radziszewski, "Private Military Companies," 795–821.

8. Avant and Neu, "The Private Security Events," 1986–2006.

9. Ibid.

10. See Isenberg, "Best of Luck to You"; Eden, "DynCorp Ends."

11. Dunigan, *Victory for Hire*, 156.

12. Abigail Fielding-Smith, Crofton Black, and the Bureau of Investigative Journalism, "A Security Company Cashed In on America's Wars—and Then Disappeared," *The Atlantic*, January 29, 2019, https://www.theatlantic.com/international/archive/2019/01/afghanistan-civilian-private-security/581263/.

13. Molly McKew, "The Gerasimov Doctrine," *Politico*, September 5, 2017, https://www.politico.eu/article/new-battles-cyberwarfare-russia/.

14. Tod Lindberg, "A U.S. Battlefield Victory against Russia's 'Little Green Men,'" *Wall Street Journal*, April 3, 2018, https://www.wsj.com/articles/a-u-s-battlefield-victory-against-russias-little-green-men-1522792572.

15. Harding and Burke, "Leaked Documents."

16. McKew, "The Gerasimov."

17. Neil Hauer, "Russia's Favorite Mercenaries," *The Atlantic*, August 27, 2018, https://www.theatlantic.com/international/archive/2018/08/russian-mercenaries-wagner-africa/568435/.

18. Carlos Ortiz, "The Market for Force in the United Kingdom: The Recasting of the Monopoly of Violence and the Management of Force as a Public-Private Enterprise," in *The Market for Force: Privatization of Security Across World Regions*, ed. Molly Dunigan and Ulrich Petersohn (Philadelphia: University of Pennsylvania Press, 2015), 58–59.

19. Hauer, "Russia's Favorite."

20. Allison, "Informal but Diverse," 87–90.

21. Hauer, "Russia's Favorite."

22. "Diyala Governorate Assessment Report," United Nations High Commissioner for Refugees, November 2016, accessed July 28, 2019, https://www.unhcr.org/462cbf672.pdf, 6.

23. Tkach, "Private Military and Security Companies, Contract Structure," 291–311.

24. Madeleine Brand, "Tactics of Iraqi Insurgents Evolving," National Public Radio, August 4, 2005, https://www.npr.org/templates/story/story.php?storyId=4785268.

25. Ronald Mutum, "Boko Haram Uses Drones, Mercenaries," *Daily Trust*, November 30, 2018, https://www.dailytrust.com.ng/boko-haram-use-drones-mercenaries.html; David Smith, "South Africa's Ageing White Mercenaries Who Helped

Turn Tide on Boko Haram," *The Guardian*, April 14, 2015, https://www.theguardian.com/world/2015/apr/14/south-africas-ageing-white-mercenaries-who-helped-turn-tide-on-boko-haram.

26. Michael K. Logan, Gina S. Ligon, and Douglas C. Derrick, "Measuring Tactical Innovation in Terrorist Attacks," *Journal of Creative Behavior* (July 6, 2019), https://onlinelibrary.wiley.com/doi/abs/10.1002/jocb.420.

27. "Factors DoD Considers When Choosing Best Value Processes Are Consistent with Guidance for Selected Acquisitions," US Government Accountability Office, July 2014, https://www.gao.gov/assets/670/665124.pdf, 1.

28. Peter Beaumont, "Abu Ghraib Abuse Firms Are Rewarded," *The Guardian*, January 15, 2015, https://www.theguardian.com/world/2005/jan/16/usa.iraq.

29. "State Department Admits No-Bid Contract 'Violates' Obama Campaign Pledges," Fox News, January 31, 2010, https://www.foxnews.com/politics/state-department-admits-no-bid-contract-violates-obama-campaign-pledges.

30. Ibid.

31. Rosie Gray, "Erik Prince's Plan to Privatize the War in Afghanistan," *The Atlantic*, August 18, 2017, https://www.theatlantic.com/politics/archive/2017/08/afghanistan-camp-david/537324/; "Afghan Government Rejects Proposals to Privatize War," Reuters, October 5, 2018, https://www.reuters.com/article/us-afghanistan-security/afghan-government-rejects-proposals-to-privatize-war-idUSKCN1MF0IH.

32. Kevin LaCroix, "Dodd Frank Anti-Retaliation Provisions Do Not Protect Overseas Whistleblowers," *The D & O Diary*, August 18, 2014, https://www.dandodiary.com/2014/08/articles/employment-practices-liability-2/dodd-frank-anti-retaliation-provisions-do-not-protect-overseas-whistleblowers/.

# Bibliography

Akcinaroglu, Seden. "Rebel Interdependencies and Civil War Outcomes." *Journal of Conflict Resolution* 56, no. 5 (2012): 879–903.

Akcinaroglu, Seden, and Elizabeth Radziszewski. "Expectations, Rivalries, and Civil War Duration." *International Interactions* 31, no. 4 (2005): 349–374.

Akcinaroglu, Seden, and Elizabeth Radziszewski. "Private Military Companies, Opportunities, and Termination of Civil Wars in Africa." *Journal of Conflict Resolution* 57, no. 5 (2013): 795–821.

Al Jazeera. "Blackwater Banned from Iraq." January 30, 1999. https://www.aljazeera.com/news/middleeast/2009/01/2009129103918814445.html.

Allison, Olivia. "Informal but Diverse: The Market for Exported Force from Russia and Ukraine." In *The Market for Force: Privatization of Security across World Regions*, edited by Molly Dunigan and Ulrich Petersohn, 87–102. Philadelphia: University of Pennsylvania Press, 2015.

Anderson, Noel. "Competitive Intervention, Protracted Conflict, and the Global Prevalence of Civil War." *International Studies Quarterly* 63, no. 3 (2019): 692–706.

Apuzzo, Matt. "Ex Blackwater Guards Given Long Terms after Killing Iraqis." *New York Times*, April 14, 2015. http://www.nytimes.com/2015/04/14/us/ex-blackwater-guards-sentenced-to-prison-in-2007-killings-of-iraqi-civilians.html.

Armendariz, Leticia. "Human Rights Abuses and Monitoring," interview by the author, Seden Akcinaroglu, January 10, 2014.

Armstrong, Matt. "Private Security Company Association of Iraq." MountainRunner. us Blog, March 29, 2006. https://mountainrunner.us/2006/03/private_security_company_assoc/.

ANSI/ASIS International. "Management System for Quality of Private Security Company Operations—Requirements with Guidance," 1–81. March 5, 2012. https://www.acq.osd.mil/log/PS/.psc.html/7_Management_System_for_Quality.pdf.

Auty, Richard. "Natural Resources and Civil Strife: A Two-Stage Process." *Geopolitics* 9, no. 1 (2004): 24–49.

Avant, Deborah. *The Market for Force: The Consequences of Privatizing Security*. New York: Oxford University Press, 2005.

Avant, Deborah. "Pragmatic Networks and Transnational Governance of Private Military and Security Services." *International Studies Quarterly* 60, no. 2 (2016): 330–342.

Avant, Deborah, and Kara Kingma Neu. "The Private Security Events Database." *Journal of Conflict Resolution* 63, no. 8 (2019): 1986–2006.

Axelrod, Alan. *Mercenaries: A Guide to Private Armies and Private Military Companies*. Thousand Oaks, CA: CQ Press, 2014.

Azad, Malyar Sadeq. "Top Leaders Tied to Security Companies." Feral-Jundi, August 21, 2010. https://feraljundi.com/afghanistan-the-pscs-connected-to-karzais-family-and-close-associates/.

Balabanova, Ekaterina. *The Media and Human Rights: The Cosmopolitan Promise.* New York: Routledge, 2015.

Balch-Lindsay, Dylan, and Andrew J. Enterline. "Killing Time: The World Politics of Civil War Duration, 1820–1992." *International Studies Quarterly* 44, no. 4 (2000): 615–642.

Barnett, Antony, and Solomon Hughes. "British Firm Accused in U.N. 'Sex Scandal.'" *The Guardian*, July 28, 2001. https://www.theguardian.com/world/2001/jul/29/unitednations.

Beaumont, Peter. "Abu Ghraib Abuse Firms Are Rewarded." *The Guardian*, January 15, 2015. https://www.theguardian.com/world/2005/jan/16/usa.iraq.

Beckhusen, Robert. "A Decade Later, Contractor Pays Out Millions for Iraq Prisoner Abuse." *Wired*, January 9, 2013. https://www.wired.com/2013/01/torture-settlement/.

Bennett, D. Scott. "Governments, Civilians, and the Evolution of Insurgency: Modeling the Early Dynamics of Insurgencies." *Journal of Artificial Societies and Simulations* 11, no. 4 (2008). http://jasss.soc.surrey.ac.uk/11/4/7.html, accessed July 15, 2019.

Bergner, Daniel. "The Other Army." *New York Times Magazine*, August 14, 2005. https://www.nytimes.com/2005/08/14/magazine/the-other-army.html.

Blair, Daniel R., and Evelyn R. Klemstine. "Afghan National Police Training Program: Lessons Learned during the Transition of Contract Administration," 1–68. A Joint Audit by the Inspectors General of Department of State and Department of Defense, August 15, 2011. https://www.hsdl.org/?abstract&did=685111.

Blaser, Noah. "Why Turkey's PKK Conflict Looms Larger Than Ever in Local Election Aftermath." Foreign Policy Research Institute, April 9, 2019. https://www.fpri.org/article/2019/04/why-turkeys-pkk-conflict-looms-larger-than-ever-in-local-election-aftermath/.

Boxell, James. "Competition Hits Security Groups in Iraq." *Financial Times*, November 5, 2005.

Brand, Madeleine. "Tactics of Iraqi Insurgents Evolving." National Public Radio, August 4, 2005. https://www.npr.org/templates/story/story.php?storyId=4785268.

Brands, Hal. "Reform, Democratization, and Counter-Insurgency: Evaluating the US Experience in Cold War–Era Latin America." *Small Wars & Insurgencies* 22, no. 2 (2011): 290–321.

British Foreign and Common Wealth Office. "Private Military Companies: Options for Regulation 2001–02," 1–48. February 12, 2002. https://assets.publishing.service.gov.uk/government/uploads/system/uploads/attachment_data/file/228598/0577.pdf.

Brooks, Doug. "Accountability and Private Contractors," interview conducted by the author, Elizabeth Radziszewski, April 2016.

Brooks, Doug. "Write a Cheque, End a War: Using Private Military Companies to End African Conflicts?," 33–35. *Conflict Trends 2000/01*, December 31, 1999. https://www.accord.org.za/publication/conflict-trends-2000-1/.

Brooks, Risa A. "The Impact of Culture, Society, Institutions, and International Forces on Military Effectiveness." In *Creating Military Power: The Sources of Military Effectiveness*, edited by Risa Brooks and Elizabeth Stanley, 1–26. Stanford, CA: Stanford University Press, 2007.

Brorsen, Les. "Looking behind the Declining Number of Public Companies." Harvard Law School Forum on Corporate Governance and Financial Regulation, May 18, 2017. https://corpgov.law.harvard.edu/2017/05/18/looking-behind-the-declining-number-of-public-companies/.

Bruce, Ian. "Charity Urges Westminster to Regulate UK Mercenary Firms." *The Herald*, February 13, 2008. https://www.heraldscotland.com/news/12749068.charity-urges-westminster-to-regulate-uk-mercenary-firms/.

Buhaug, Halvard, and Scott Gates. "The Geography of Civil War." *Journal of Peace Research* 39, no. 4 (2002): 417–433.

The Business Ethics Blog. "Security Contractor Boosts Ethics Practices." July 29, 2009. https://businessethicsblog.com/2009/07/1.

Business Wire. "CACI Wins $415 Million Contract to Develop and Deploy Intelligence Systems for U.S. Army." July 1, 2019. https://www.businesswire.com/news/home/20190509005154/en/CACI-Wins-415-Million-Contract-Develop-Deploy.

Cambanis, Thanassis. "The Logic of Assad's Brutality." *The Atlantic*, April 8, 2018. https://www.theatlantic.com/international/archive/2018/04/syria-chemical-weapons-assad-trump/557483/.

Campbell, John. "Boko Haram Conflict Enters Counterinsurgency Phase as Nigeria Erects 'Fortresses.'" Council on Foreign Relations, December 7, 2017. https://www.cfr.org/blog/boko-haram-conflict-enters-counterinsurgency-phase-nigeria-erects-fortresses.

Canetti, Daphna, and Miriam Lindner. "Exposure to Political Violence and Political Behavior." In *Psychology of Change: Life Contexts, Experiences, and Identities*, edited by Katherine Reynolds and Nyla Branscombe, 77–94. New York: Psychology Press, 2014.

Caparini, Marina. "Domestic Regulation: Licensing Regimes for the Export of Military Goods and Services." In *From Mercenaries to Market: The Rise and Regulation of Private Military Companies*, edited by Simon Chesterman and Chia Lehnardt, 158–181. New York: Oxford University Press, 2007.

Carmola, Kateri. "PMSCs and Risk in Counterinsurgency Warfare." In *Contractors and Wars: The Transformation of US Expeditionary Operations*, edited by Christopher Kinsey and Malcolm Hugh Patterson, 134–157. Stanford, CA: Stanford University Press, 2012.

Catallo, Jennifer. "China's Managed Market for Force." In *The Market for Force: Privatization of Security across World Regions*, edited by Molly Dunigan and Ulrich Petersohn, 118–132. Philadelphia: University of Pennsylvania Press, 2015.

Censer, Marjorie. "After Refocusing, DynCorp International Sees Profits Rise, Weighs Acquisitions." *Inside Defense*, May 29, 2018. https://insidedefense.com/share/196249.

Center for Systemic Peace Research. "Polity IV Project: Political Regime Characteristics and Transitions, 1800–2013." Last modified June 6, 2014. https://www.systemicpeace.org/polity/polity4.htm.

Chajet, Clive. "Corporate Reputation as a Strategic Asset: Corporate Reputation and the Bottom Line." *Corporate Reputation Review* 1 (1997): 19–22.

Chakrabarti, Shantanu. "Privatization of Security in the Post–Cold War Period," 1–75. Institute for Defense Studies and Analyses, December 2009. https://idsa.in/system/files/Monograph_No2.pdf.

Christiansesn, Anne Mette, Thomas Trier Hansen, and Riikka Poukka. "Nordic Human Rights Study," 1–27. Deloitte, 2015. https://www2.deloitte.com/content/dam/Deloitte/no/Documents/risk/Nordic-Human-Rights-Study.pdf.

Cockayne, James. "Make or Buy? Principal-Agent Theory and the Regulation of Private Military Companies." In *From Mercenaries to Market: The Rise and Regulation of Private Military Companies*, edited by Simon Chesterman and Chia Lehnardt, 196–217. New York: Oxford University Press, 2007.

Collier, Paul, and Anke Hoeffler. "Data Issues in the Study of Conflict," 1–14. Conference Paper, June 6, 2001. https://pdfs.semanticscholar.org/e777/8066cc0569d63bac254842 7b7630e6e2255b.pdf.

Collier, Paul, and Anke Hoeffler. "Greed and Grievance in Civil War." *Oxford Economic Papers* 56, no. 4 (2004): 663–695.

Collier, Paul, Anke Hoeffler, and Måns Söderbom. "On the Duration of Civil War." *Journal of Peace Research* 41 no. 3 (2004): 253–273.

Committee on Homeland Security and Governmental Affairs, US Senate. "An Uneasy Relationship: U.S. Reliance on Private Security Firms in Overseas Operations," 1–162. February 27, 2008. https://fas.org/irp/congress/2008_hr/psc.pdf.

Committee on Homeland Security and Governmental Affairs, US Senate. "Iraq Reconstruction: Lessons Learned in Contracting." 1–62. August 2006. https://www.govinfo.gov/content/pkg/CHRG-109shrg29761/html/CHRG-109shrg29761.htm.

Cotton, Sarah K., Ulrich Petersohn, Molly Dunigan, Q Burkhart, Megan Zander-Cotugno, Edward O'Connell, and Michael Webber. *Hired Guns: Views about Armed Contractors in Operation Iraqi Freedom.* Santa Monica, CA: RAND, 2010.

Cunningham, David E. "Veto Players and Civil War Duration." *American Journal of Political Science* 50, no. 4 (2006): 875–892.

Cunningham, David E., Kristian Skrede Gleditsch, and Idean Salehyan. "It Takes Two: A Dyadic Analysis of Civil War Duration and Outcome." *Journal of Conflict Resolution* 53, no. 2: 570–597.

Cusumano, Eugenio. "Regulating Private Military and Security Companies: A Multifaceted and Multilayered Approach," 1–24. EUI Working Paper, Academy of European Law, 2009.

Dalenga, Aaron W., and Brendan W. McGarry. "Defense Primer: DOD Transfer and Reprogramming Authorities." Congressional Research Service, June 7, 2019. https://fas.org/sgp/crs/natsec/IF11243.pdf.

Data on Armed Conflict and Security. "Private Security Database." https://www.conflict-data.org/psd/index.html, accessed May 5, 2019.

Demmers, Jolle. "Diaspora and Conflict: Locality, Long-Distance Nationalism, and Delocalisation of Conflict Dynamics." *Javnost—The Public* 9, no. 1 (2002): 85–96.

de Nevers, Renée. "Private Military Contractors and Changing Norms for the Laws of Armed Conflict." In *New Battlefields/Old Laws: Critical Debates on Asymmetric Warfare,* edited by William C. Banks, 150–168. New York: Columbia University Press, 2011.

*Deutsche Welle.* "German Arms Firm Ends Blackwater Deal after TV Report." February 19, 2008. https://www.dw.com/en/german-arms-firm-ends-blackwater-deal-after-tv-report/a-3135177.

Dickinson, Laura. "Contract as a Tool for Regulating Private Military Companies." In *From Mercenaries to Market: The Rise and Regulation of Private Military Companies,* edited by Simon Chesterman and Chia Lehnardt, 217–239. New York: Oxford University Press, 2007.

Dixon, Jeffrey S., and Meredith Reid Sarkees. *A Guide to Intra-State Wars: An Examination of Civil, Regional, and Intercommunal Wars, 1816–2007.* Washington, DC: CQ Press, 2016.

Donelson, Dain, Justin J. Hopkins, and Christopher G. Yust. "The Cost of Disclosure Regulation: Evidence from D & O Insurance and Non-Meritorious Securities

Litigation." The Columbia Law School Blue Sky Blog, February 22, 2018. http://clsbluesky.law.columbia.edu/2018/02/22/the-cost-of-disclosure-regulation-evidence-from-do-insurance-and-non-meritorious-securities-litigation/.

Drutschmann, Sebastian. "Informal Regulation: An Economic Perspective on the Private Security Industry." In *Private Military and Security Companies: Chances, Problems, Pitfalls, and Prospects*, edited by Thomas Jager and Gerhard Kummel, 443–455. Wiesbaden: VS Verlag für Sozialwissenschaften, 2007.

Drutschmann, Sebastian. "Motivation, Markets and Client Relations in the British Private Security Industry." PhD diss., King's College, 2014.

Dunar, Charles J., III, Jared J. Mitchell, and Donald L. Robbins III. "Private Military Industry Analysis: Private and Public Companies." MBA thesis, Naval Postgraduate School, 2007.

Dunigan, Molly. *Victory for Hire: Private Security Companies' Impact on Military Effectiveness*. Stanford, CA: Stanford University Press, 2011.

Eden, Scott. "DynCorp Ends Public Experiment." *The Street*, April 13, 2010. https://www.thestreet.com/investing/stocks/dyncorp-ends-public-experiment-10723699.

Editorial Board, "Runaway Spending on War Contractors." *New York Times*, September 17, 2001. https://www.nytimes.com/2011/09/18/opinion/sunday/runaway-spending-on-war-contractors.html.

Eland, Ivan. *The Failure of Counterinsurgency: Why Hearts and Minds Are Not Always Won*. Santa Barbara, CA: Praeger, 2013.

Elbadawi, Ibrahim A., and Nicholas Sambanis. "External Interventions and the Duration of Civil Wars," 1–18. World Bank Working Paper, September 1, 2000. http://documents.worldbank.org/curated/en/760801468766521682/External-interventions-and-the-duration-of-civil-wars.

Elms, Heather, and Robert A. Phillips. "Private Security Companies and Institutional Legitimacy: Corporate and Stakeholder Responsibility." *Business Ethics Quarterly* 19, no. 3 (2009): 403–432.

Enders, Walter, Todd Sandler, and Khusrav Gaibulloev. "Domestic versus Transnational Terrorism: Data, Decomposition, and Dynamics." *Journal of Peace Research* 48, no. 3 (2011): 319–337.

Erikson, Erik Daniel. "Meeting ISO 18788 Criteria." International Foundation for Protection Officers, June 2018. http://www.ifpo.org/wp-content/uploads/2018/06/18788_EDE_article.doc.

Fainaru, Steve. "Cutting Costs, Bending Rules, and a Trail of Broken Lives." *Washington Post*, July 29, 2007. https://www.washingtonpost.com/wp-dyn/content/article/2007/07/28/AR2007072801407.html.

Fainaru, Steve. "Iraq Contracts Face Growing Parallel War: As Security Work Increases So Do the Casualties." *Washington Post*, June 16, 2007. https://www.washingtonpost.com/wp-dyn/content/article/2007/06/15/AR2007061502602_3.html.

Fainaru, Steve. "Where Military Rules Don't Apply: Blackwater's Security Force in Iraq Given Wide Latitude by State Dept." *Washington Post*, September 20, 2007. https://www.washingtonpost.com/wp-dyn/content/article/2007/09/19/AR2007091902503_4.html?sid=ST2007092000478.

Fearon, James D. "Why Do Some Civil Wars Last So Much Longer Than Others?" *Journal of Peace Research* 41, no. 3 (2004): 275–301.

Fearon, James D., and David D. Laitin. "Ethnicity, Insurgency and Civil War. *American Political Science Review* 97, no. 1 (2003): 75–90.

Federation of European Accountants. "Integrity in Professional Ethics." September 2009. https://www.icaew.com/-/media/corporate/files/technical/ethics/integrity-in-professional-ethics-fee-discussion-paper.ashx?la=en.

Felter, Joseph, and Jake Shapiro. "Limiting Civilian Casualties as Part of a Winning Strategy: The Case of Courageous Restraint." *Daedalus* 146, no. 1 (2017): 48–50.

Fielding-Smith, Abigail, Crofton Black, and the Bureau of Investigative Journalism. "A Security Company Cashed In on America's Wars—and Then Disappeared." *The Atlantic*, January 29, 2019. https://www.theatlantic.com/international/archive/2019/01/afghanistan-civilian-private-security/581263/.

Finel, Bernard. "An Alternative to COIN." *Armed Forces Journal*, February 1, 2010. http://armedforcesjournal.com/an-alternative-to-coin/.

Fitzsimmons, Scott. *Mercenaries in Asymmetric Conflicts*. New York: Cambridge University Press, 2003.

Fitzsimmons, Scott. *Private Security Companies during the Iraq War: Military Performance and the Use of Deadly Force*. New York: Routledge, 2016.

Fombrun, Charles. *Reputation: Realizing Value from the Corporate Image*. Boston: Harvard Business School Press, 1996.

Former Blackwater Employee, "Private Security Accountability," interview by the author, Elizabeth Radziszewski, August 26, 2012.

Former Pacific Architects and Engineers employee. "Competition and Effectiveness," interview by the author, Elizabeth Radziszewski, April 29, 2012.

Former Pacific Architects and Engineers employee. "Private Military Companies," interview by the author, Elizabeth Radziszewski, April 10, 2012.

Former PMSC employee, "Accountability and Private Contractors," interview by the author, Elizabeth Radziszewski, September 28, 2012.

Fortna, Virginia Page. "Do Terrorists Win? Rebels' Use of Terrorism and Civil War Outcomes." *International Organization* 69, no. 3 (2015): 519–556.

Fox News. "State Department Admits No-Bid Contract 'Violates' Obama Campaign Pledges." January 31, 2010. https://www.foxnews.com/politics/state-department-admits-no-bid-contract-violates-obama-campaign-pledges.

Francis, David J. "Mercenary Intervention in Sierra Leone: Providing National Security or International Exploitation?" *Third World Quarterly* 20, no.2 (1999): 319–338.

G4S. "G4S 2017 Full Year Results," 1–37. March 8, 2018. https://www.g4s.com/-/media/g4s/corporate/files/investor-relations/2017/prelim2017fullyearresults08032018.ashx?la=en&hash=0DD149ECC59490855C37772F70BA6A11.

Glass, Charles. "The Warrior Class: A Golden Age for the Freelance Soldier." *Harper's Magazine*, April 2012. https://harpers.org/archive/2012/04/the-warrior-class/.

Gleditsch, Nils Petter, Peter Wallensteen, Mikael Eriksson, Margareta Sollenberg, and Håvard Strand. "Armed Conflict 1946–2001: A New Dataset." *Journal of Peace Research* 39 no. 5 (2002): 625–637.

Gomez del Prado, Jose. "The Privatization of War: Mercenaries, Private Military and Security Companies (PMSC)." Global Research, April 9, 2016. https://www.globalresearch.ca/the-privatization-of-war-mercenaries-private-military-and-security-companies-pmsc/21826.

Gray, Rosie. "Erik Prince's Plan to Privatize the War in Afghanistan." *The Atlantic*, August 18, 2017. https://www.theatlantic.com/politics/archive/2017/08/afghanistan-camp-david/537324/.

Griffing, Alexander. "How Assad Helped Create ISIS to Win in Syria and Got Away with the Crime of the Century." *Haaretz*, October 7, 2018. https://www.haaretz.com/middle-east-news/syria/MAGAZINE-iran-russia-and-isis-how-assad-won-in-syria-1.6462751.

Griffiths, Robert J. *U.S. Security Cooperation with Africa: Political and Policy Challenges.* New York: Routledge, 2016.

Gulam, Hyder. "The Rise and Rise of Private Military Companies." Master's thesis, Peace Operations Training Institute, 2005.

Gutiérrez-Sanin, Francisco. "Criminal Rebels? Discussion of Civil War and Criminality from the Colombian Experience." *Politics & Society* 32, no. 2 (2004): 257–285.

Haeberle, Kevin, and M. Todd Henderson. "Making a Market for Corporate Disclosure." *Yale Journal of Regulation* 35, no. 2 (2018): 383–436.

Hanson, Dallas, and Helen Stuart. "Failing the Reputation Management Test: The Case of BHP, the Big Australian." *Corporate Reputation Review* 4, no. 2 (2001):128–143.

Harding, Luke, and Jason Burke. "Leaked Documents Reveal Russian Effort to Exert Influence in Africa." *The Guardian*, June 11, 2019. https://www.theguardian.com/world/2019/jun/11/leaked-documents-reveal-russian-effort-to-exert-influence-in-africa.

Hauer, Neil. "Russia's Favorite Mercenaries." *The Atlantic*, August 27, 2018. https://www.theatlantic.com/international/archive/2018/08/russian-mercenaries-wagner-africa/568435/.

Herbst, Jeffrey. "African Militaries and Rebellion: The Political Economy of Threat and Combat Effectiveness." *Journal of Peace Research* 41, no. 3 (2004): 357–369.

Herbst, Jeffrey. "The Regulation of Private Security Forces." In *The Privatization of Security in Africa*, edited by Greg Mills and John Stremlau, 107–129. Johannesburg: SAIIA Press, 1999.

Heston, Alan, Robert Summers, and Bettina Aten. "Penn World Table Version 6.3." Center for International Comparison of Production Income, and Prices, University of Pennsylvania, August 2009. http://datacentre2.chass.utoronto.ca/pwt/, accessed June 8, 2013.

Hibou, Beatrice. "From Privatizing the Economy to Privatizing the State: An Analysis of the Continual Formation of the State." In *Privatizing the State*, edited by Beatrice Hibou, 1–48. New York: Columbia University Press, 2004.

Hinton, Christopher. "DynCorp to Go Private in $1.5 Billion Deal with Cerberus." Market Watch, April 12, 2010. https://www.marketwatch.com/story/cerberus-to-take-dyncorp-private-for-15-billion-2010-04-12.

Houbert, Jean, "The Mascareignes, the Seychelles and the Chagos, Islands with a French Connection: Security in a Decolonized Indian Ocean." In *Political Economy of Small Tropical Islands: The Importance of Being Small*, edited by Helen M. Hintjens and Marilyn D. D. Newitt, 93–112. Exeter, UK: University of Exeter Press, 1992.

Howe, Herbert. "Private Security Forces and African Stability: The Case of Executive Outcomes." *Journal of Modern African Studies* 36, no. 2 (1998): 307–331.

Hsu, Spencer. "Blackwater Guard's Retrial for Murder in 2007 Iraq Massacre Goes to U.S. Jury." *Washington Post*, August 8, 2018. https://www.washingtonpost.com/local/public-safety/blackwater-guards-retrial-for-murder-in-2007-iraq-massacre-goes-to-us-jury/2018/08/08/710671a8-98cd-11e8-843b-36e177f3081c_story.html.

Hultquist, Phillip. "Power Parity and Peace? The Role of Relative Power in Civil War Settlement." *Journal of Peace Research* 50, no. 5 (2013): 623–634.

Humphreys, Macartan, and Jeremy M. Weinstein. "Who Fights? The Determinants of Participation in Civil War." *American Journal of Political Science* 52, no. 2 (2008): 436–455.

International Code of Conduct Association. "International Code of Conduct for Private Security Service Providers," 1–15. November 9, 2010. https://www.icoca.ch/sites/all/themes/icoca/assets/icoc_english3.pdf.

International Code of Conduct Association. "Procedures Article 12: Reporting, Monitoring and Assessing Performance and Compliance," 1–8. https://www.icoca.ch/sites/default/files/uploads/ICoCA-Procedures-Article-12-Monitoring.pdf, accessed May 12, 2016.

International Committee of the Red Cross. "The Geneva Conventions of August 12, 1949." August 12, 1949. https://ihl-databases.icrc.org/ihl/INTRO/380, 153–221.

International Committee of the Red Cross. "Protocol Additional to the Geneva Conventions of 12 August 1949, and Relating to the Protection of Victims of International Armed Conflicts (Protocol 1)." June 8, 1977. https://ihl-databases.icrc.org/ihl/WebART/470-750057.

International Committee of the Red Cross and Swiss FDFA. "The Montreux Document: On Pertinent International Legal Obligations and Good Practices for States Related to Operations of Private Military and Security Companies during Armed Conflict," 7–27. September 17, 2008. https://www.icrc.org/en/doc/assets/files/other/icrc_002_0996.pdf.

International Peace Research Institute (PRIO). "UCDP/PRIO Armed Conflict Dataset Codebook Version 4–2009," 1–16. https://www.prio.org/Global/upload/CSCW/Data/UCDP/2009/Codebook_UCDP_PRIO%20Armed%20Conflict%20Dataset%20v4_2009.pdf, accessed May 10, 2019.

International Stability Operations Association. "ISOA Code of Conduct." April 1, 2001. https://stability-operations.org/page/Code.

Isenberg, David. "Best of Luck to You Ms. Burke." CATO Institute, May 14, 2010. http://www.cato.org/publications/commentary/best-luck-you-ms-burke-0.

Isenberg, David. "PMC = Private Military Costs." *HuffPost*, January 28, 2013. http://www.huffingtonpost.com/david-isenberg/pmc-private-military-cost_b_2208825.html.

Isenberg, David. "Private Military Contractors and U.S. Grand Strategy," 5–47. International Peace Research Institute (PRIO) Report, 2009. https://www.files.ethz.ch/isn/109297/Isenberg%20Private%20Military%20Contractors%20PRIO%20Report%201-2009.pdf.

ISO (The International Organization for Standardization). "Management System for Private Security Operations—Requirements with Guidance for Use," 1–98. September 2015. https://www.iso.org/standard/63380.html.

Jägers, Nicole. "Will Transnational Private Regulation Close the Governance Gap?" In *Human Rights Obligations of Business: Beyond the Corporate Responsibility to Protect?*, edited by Surya Deva and David Bilchitz, 295–329. New York: Cambridge University Press, 2013.

Joachim, Jutta, Marlene Martin, Henriette Lange, Andrea Schneiker, and Magnus Dau. "Twittering for Talent: Private Military and Security Companies between Business and Military Branding." *Contemporary Security Policy* 39, no. 2 (2018): 298–316.

Kalyvas, Stathis. "'New' and 'Old' Civil Wars: A Valid Distinction?" *World Politics* 54, no. 1 (2001): 99–118.

Kalyvas, Stathis, and Matthew Kocher. "How 'Free' Is Freeriding in Civil Wars?: Violence, Insurgency, and the Collective Action Problem." *World Politics* 59, no. 2 (2007): 177–216.

Kathman, Jacob D. "Civil War Diffusion and Regional Motivations for Intervention." *Journal of Conflict Resolution* 55, no. 6 (2011): 847–876.

Kaufmann, Chaim. "Possible and Impossible Solutions to Ethnic Civil Wars." *International Security* 20, no. 4 (1996): 136–175.

Kennedy, Kevin C. "A Critical Appraisal of Criminal Deterrence Theory. *Dickinson Law Review* 88, no. 1 (1983–1984): 1–14.

Kilcullen, David. "Twenty-Eight Articles: Fundamentals of Company-Level Counterinsurgency," 29–35. IO Sphere Joint Information Operations Center, summer 2006. https://www.pegc.us/archive/Journals/iosphere_summer06_kilcullen.pdf.

Kilcullen, David. "Two Schools of Classic Counterinsurgency." *Small Wars Journal* Blog, January 27, 2007. http://smallwarsjournal.com/blog/two-schools-of-classical-counterinsurgency.

Kinsey, Christopher. *Corporate Soldiers and International Security: The Rise of Private Military Companies.* New York: Routledge, 2006.

Kinsey, Christopher. "Private Security Companies and Corporate Social Responsibility." In *Private Military and Security Companies: Ethics, Policies and Civil-Military Relations*, edited by Andrew Alexandra, Deane-Peter Baker and Marina Caparini, 70–86. New York, NY: Routledge, 2008.

Kinsey, Christopher. *Private Contractors and the Reconstruction of Iraq: Transforming Military Logistics.* New York: Routledge, 2009.

Kirschner, Shanna A. "Families and Foes: Ethnic Civil War Duration." PhD diss., University of Michigan, 2009.

Kleck, Gary, Brion Sever, Spencer Li, and Marc Gertz. "The Missing Link in General Deterrence Research." *Criminology* 43, no. 3 (2005): 623–659.

Klein, Alex. "U.S. Army Awards Iraq Security Work to British Firm." *Washington Post*, September 14, 2007. http://www.washingtonpost.com/wp-dyn/content/article/2007/09/13/AR2007091302237.html.

Kouzes, James, and Barry Posner. *The Leadership Challenge.* San Francisco: Jossey Bass, 2002.

Krahmann, Elke. "Choice, Voice and Exit: Consumer Power and the Self-Regulation of the Private Security Industry." *European Journal of International Security* 1, no. 1 (2016): 27–48.

Krahmann, Elke. "NATO Contracting in Afghanistan: The Problem of Principal-Agent Networks." *International Affairs* 92, no. 6 (2016): 1401–1426.

Labott, Elise. "U.S. Will Not Renew Iraq Contract with Blackwater." CNN, January 30, 2009. http://edition.cnn.com/2009/WORLD/meast/01/30/us.blackwater.contract/.

Lacina, Bethany, and Nils Petter Gleditsch. "Monitoring Trends in Global Combat: A New Dataset of Battle Deaths." *European Journal of Population* 21, no. 2–3 (2005): 145–166.

LaCroix, Kevin. "Dodd Frank Anti-Retaliation Provisions Do Not Protect Overseas Whistleblowers." *The D & O Diary*, August 18, 2014. https://www.dandodiary.com/2014/08/articles/employment-practices-liability-2/dodd-frank-anti-retaliation-provisions-do-not-protect-overseas-whistleblowers/.

Leander, Anna. "The Market for Force and Public Security: The Destabilizing Consequences of Private Military Companies." *Journal of Peace Research* 42, no. 5 (2005): 605–622.

Leander, Anna. "The Power to Construct International Security: On the Significance of Private Military Companies." *Millennium: Journal of International Studies* 33, no. 3 (2005): 803–825.

LeBillon, Phillippe. "The Political Ecology of War: Natural Resources and Armed Conflicts." *Political Geography* 20, no. 5 (2001): 561–584.

Lindberg, Tod. "A U.S. Battlefield Victory against Russia's 'Little Green Men.'" *Wall Street Journal*, April 3, 2018. https://www.wsj.com/articles/a-u-s-battlefield-victory-against-russias-little-green-men-1522792572.

Lindblom, Lars. "Dissolving the Moral Dilemma of Whistleblowing." *Journal of Business Ethics* 76, no. 4 (2007): 413–426.

Lindorff, Dave. "Exclusive: The Pentagon's Massive Fraud Exposed." *The Nation*, January 7, 2019. https://www.thenation.com/article/pentagon-audit-budget-fraud/.

Logan, Michael K., Gina S. Ligon, and Douglas C. Derrick. "Measuring Tactical Innovation in Terrorist Attacks." *Journal of Creative Behavior*, July 6, 2019. https://onlinelibrary.wiley.com/doi/abs/10.1002/jocb.420.

Lowenheim, Nava. "Turkey's Dual Problem: Between Armenia and the Armenian Diaspora." In *Nonstate Actors in Intrastate Conflicts*, edited by Dan Miodownik and Oren Barak, 106–125. Philadelphia: University of Pennsylvania Press, 2014.

Luttwak, Edward. "Dead End: Counterinsurgency Warfare as Military Malpractice." *Harper's Magazine*, February, 2007. https://harpers.org/archive/2007/02/dead-end/.

Lyall, Jason. "Civilian Casualties, Radicalization, and the Effects of Humanitarian Assistance in Wartime Settings," 1–38. Unpublished manuscript, April 5, 2015. http://aiddata.org/sites/default/files/lyall_2015_humanitarian_aid_afghanistan.pdf.

Lyall, Jason, and Isaiah Wilson III. "Rage against the Machines: Explaining Outcomes in Counterinsurgency Wars." *International Organization* 63, no. 1 (2009): 67–106.

Macartan, Humphreys. "Natural Resources, Conflict, and Conflict Resolution: Uncovering the Mechanisms." *Journal of Conflict Resolution* 49, no. 4 (2005): 508–537.

Macleod, Sorcha, and Rebecca Dewinter-Schmitt. "Certifying Private Security Companies: Effectively Ensuring the Corporate Responsibility to Respect Human Rights?" *Business and Human Rights Journal* 4, no. 1 (2019): 55–77.

Mani, Kristina. "Diverse Markets for Force in Latin America: From Argentina to Guatemala." In *The Market for Force: Privatization of Security across World Regions*, edited by Molly Dunigan and Ulrich Petersohn, 20–38. Philadelphia: University of Pennsylvania Press, 2015.

Market Watch. "DynCorp to Go Private in $1.5 Billion Deal with Cerberus." April 12, 2010. https://www.marketwatch.com/story/cerberus-to-take-dyncorp-private-for-15-billion-2010-04-12.

Mason, T. David, and Dale A. Krane. "The Political Economy of Death Squads: Toward a Theory of the Impact of State-Sanctioned Terror." *International Studies Quarterly* 33, no. 2 (1989): 175–198.

Mason, T. David, Joseph P. Weingarten, and Patrick J. Fett. "Win, Lose, or Draw: Predicting the Outcome of Civil Wars." *Political Research Quarterly* 52, no. 2 (1999): 239–268.

Masser, Barbara, and Rupert Brown. "When Would You Do It? An Investigation into the Effects of Retaliation, Seriousness of Malpractice and Occupation on Willingness to Blow the Whistle." *Journal of Community & Applied Social Psychology* 6, no. 2 (1996): 127–130.

Mathieu, Fabien, and Nick Dearden. "Corporate Mercenaries: The Threat of Private Military & Security Companies." *Review of African Political Economy* 34, no. 114 (2007): 744–755.

Mayer, Christopher. "DoD Monitoring," interview by the author, Elizabeth Radziszewski, August 5, 2019.

Mayer, Christopher. "Private Security, Military Contractors." Interview by the author, Elizabeth Radziszewski, January 19, 2019.

Mayer, Don. "Peaceful Warriors: Private Military Security Companies and the Quest for Stable Societies." *Journal of Business Ethics* 89, no. 4 (2009): 387–401.

McCarthy, Libby. "New Report Reveals 86% of U.S. Consumers Expect Companies to Act on Social, Environmental Issues." Sustainable Brands, 2017. https://sustainablebrands. com/read/marketing-and-comms/new-report-reveals-86-of-us-consumers-expect-companies-to-act-on-social-environmental-issues, accessed April 5, 2019.

McFate, Sean. "America's Addiction to Mercenaries." *The Atlantic*, August 12, 2016. https://www.theatlantic.com/international/archive/2016/08/iraq-afghanistan-contractor-pentagon-obama/495731/.

McFate, Sean. *The Modern Mercenary*. New York: Oxford, 2014.

McIntyre, Angela, and Taya Weiss. "Weak Governments in Search of Strength: Africa's Experience of Mercenaries and Private Military Companies." In *From Mercenaries to Market: The Rise and Regulation of Private Military Companies*, edited by Simon Chesterman and Chia Lehnardt, 67–82. New York: Oxford University Press, 2007.

McKew, Molly. "The Gerasimov Doctrine." *Politico*, September 5, 2017. https://www.po-litico.eu/article/new-battles-cyberwarfare-russia/.

Menocal, Alina Rocha, Verna Fritz, and Lisa Rakner. "Hybrid Regimes and the Challenges of Deepening and Sustaining Democracy in Developing Countries." *South African Journal of International Affairs* 15, no. 1 (2008): 29–40.

Millett, Allan R., Williamson Murray, and Kenneth H. Watman. "The Effectiveness of Military Organization." *International Security* 11, no. 1 (1986): 37–71.

Millson, Christopher. "Comparing Counterinsurgency Tactics in Iraq and Vietnam." *Inquiries Journal* 3, no. 5 (2011): 1.

Miodownik, Dan, Oren Barak, Maayan Mor, and Omer Yair. "Introduction." In *Nonstate Actors in Intrastate Conflicts*, edited by Dan Miodownik and Oren Barak, 1–12. Philadelphia: University of Pennsylvania Press, 2014.

Mulgan, Richard. "Comparing Accountability in the Public and Private Sectors." *Australian Journal of Public Administration* 59, no. 1 (2000): 87.

Musah, Abdel-Fatau, and J. Kayode Fayemi. "Africa in Search of Security: Mercenaries and Conflicts—An Overview." In *Mercenaries: An African Security Dilemma*, edited by Abdel-Fatau Musah and J. Kayode Fayemi, 13–42. Sterling, VA: Pluto, 2000.

Mutum, Ronald. "Boko Haram Uses Drones, Mercenaries." *Daily Trust*, November 30, 2018. https://www.dailytrust.com.ng/boko-haram-use-drones-mercenaries.html.

Newell, Virginia, and Benedict Sheehy. "Corporate Militaries and States: Actors, Interactions, and Reactions." *Texas International Law Journal* 41, no. 1 (2006): 67–102.

Norton-Taylor, Richard, and Richard Wray. "Boss Quits ArmorGroup after Iraq Problems." *The Guardian*, November 28, 2007. https://www.theguardian.com/business/2007/nov/28/iraq.

Nossiter, Adam. "After Vote in Congo, Talk of Resistance." *New York Times*, December 7, 2011. https://www.nytimes.com/2011/12/08/world/africa/after-vote-in-congo-talk-of-resistance.html?mtrref=www.google.com&gwh=C6E700DF52B47DD302CE73A3 9BB94DA5&gwt=pay&assetType=REGIWALL.

O'Brien, James M., III. "Private Military Companies: An Assessment." Master's thesis, Naval Postgraduate School, 2008.

O'Brien, Kevin. "Private Military Companies and African Security, 1990–8." In *Mercenaries: An African Security Dilemma*, edited by Abdel-Fatau Musah and J. Kayode Fayemi, 43–75. London: Pluto, 2000.

O'Brien, Kevin. "What Should and What Should Not Be Regulated." In *From Mercenaries to Market: The Rise and Regulation of Private Military Companies*, edited by Simon Chesterman and Chia Lehnardt, 29–49. New York: Oxford University Press, 2007.

Office of Inspector General. "Performance Evaluation of PAE Operations and Maintenance Support for the Bureau of International Narcotics and Law Enforcement Affairs' Counternarcotics Compounds in Afghanistan," 1–35. February 2011. https://books.google.com/books?id=emQ4YKGANSYC&pg=PA7&lpg=PA7&dq=Performance+Evaluation+of+PAE+Operations+2011&source=bl&ots=eYJiItl4NH&sig=ACfU3U2Hb4PaWjV4NkaPDU35KWuSLAYTcQ&hl=en&ppis=_c&sa=X&ved=2ahUKEwjn8aHt4N7mAhVvRN8KHb0RDrkQ6AEwAXoECAoQAQ#v=onepage&q=Performance%20Evaluation%20of%20PAE%20Operations%202011&f=false.

Olasupo, Abisola. "Nigeria: How Buhari Stopped Us from Fighting Boko Haram—South African Mercenary." *The Guardian*, November 26, 2018. https://allafrica.com/stories/201811270024.html.

Olonisakin, Funmi. "Arresting the Tide of Mercenaries: Prospects for Regional Control." In *Mercenaries: An African Security Dilemma*, edited by Abdel-Fatau Musah and J. Kayode Fayemi, 233–257. Sterling, VA: Pluto, 2000.

Ortiz, Carlos. "The Market for Force in the United Kingdom: The Recasting of the Monopoly of Violence and the Management of Force as a Public-Private Enterprise." In *The Market for Force: Privatization of Security across World Regions*, edited by Molly Dunigan and Ulrich Petersohn, 52–71. Philadelphia: University of Pennsylvania Press, 2015.

Osorio, Javier. "Democratization and Drug Violence in Mexico," 1–69. Working Paper. https://eventos.itam.mx/sites/default/files/eventositammx/eventos/aadjuntos/2014/01/democratizacion_and_drug_violence_osorio_appendix_1.pdf, accessed October 29, 2018.

Ouedraogo, Emile. "Advancing Military Professionalism in Africa," 1–53. African Center for Strategic Studies, July 2014. https://africacenter.org/wp-content/uploads/2016/06/ARP06EN-Advancing-Military-Professionalism-in-Africa.pdf.

Palou-Loverdos, Jordi, and Leticia Armendariz. "The Privatization of Warfare, Violence and Private Military & Security Companies: A Factual and Legal Approach to Human Rights Abuses by PMSC in Iraq," 1–332. Institute for Active Nonviolence Action, 2011. http://www.consciousbeingalliance.com/Informe_PMSC_Iraq_Nova.pdf.

Paul, Christopher, Colin P. Clarke, and Beth Grill. *Victory Has a Thousand Words: Sources of Success in Counterinsurgency*. Santa Monica, CA: RAND, 2010.

Pech, Khareen. "Executive Outcomes—A Corporate Conquest." In *Peace, Profit or Plunder? The Privatization of Security in War-Torn African Societies*, edited by Jakkie Cilliers and Douglas Fraser, 81–110. Pretoria, South Africa: Institute for Security Studies, 1999.

Peters, Heidi, and Sofia Plagakis. "Department of Defense Contractor and Troop Levels in Afghanistan and Iraq: 2007–2018," 1–18. Congressional Research Service, May 10, 2019. https://fas.org/sgp/crs/natsec/R44116.pdf.

Petersohn, Ulrich. "The Effectiveness of Contracted Coalitions: Private Security Contractors in Iraq." *Armed Forces & Society* 39, no. 3 (2013): 467–488.

Petersohn, Ulrich. "The Impact of Mercenaries and Private Military and Security Companies on Civil War Severity between 1946 and 2002." *International Interactions* 40, no. 2 (2014): 191–215.

Petersohn, Ulrich. "Private Military and Security Companies (PMSCs), Military Effectiveness, and Conflict Severity in Weak States, 1990–2007." *Journal of Conflict Resolution* 61, no. 5 (2017): 1046–1072.

Petersohn, Ulrich. "The Social Structure of the Market for Force." *Cooperation and Conflict* 50, no. 2 (2015): 269–285.

Petersohn, Ulrich, and Molly Dunigan. "Introductioin." In *The Market for Force: Privatization of Security across World Regions*, edited by Molly Dunigan and Ulrich Petersohn, 1–20, Philadelphia: University of Pennsylvania Press, 2015.

Peterson, Laura, and Phillip van Niekerk. "Privatizing Combat—The New World Order." The Public i-Center for Public Integrity, November/December 2002. http://www.thirdworldtraveler.com/War_Peace/Privatizing_Combat.html.

Pettersson, Therese, Stina Hogbladh, and Magnus Oberg. "Organized Violence, 1989-2018 and Peace Agreements." *Journal of Peace Research* 56, no. 4 (2019): 589–603.

Pleming, Sue, and Andy Sullivan. "FBI Takes Lead in Blackwater Investigation." Reuters, October 4, 2007. https://www.reuters.com/article/us-usa-iraq-blackwater/fbi-takes-lead-in-blackwater-investigation-idUSN0430861520071005.

Porter, Michael. "The Five Competitive Forces That Shape Strategy." *Harvard Business Review*, January 2008. https://hbr.org/2008/01/the-five-competitive-forces-that-shape-strategy.

Prem, Berenike. *Private Military and Security Companies as Legitimate Governors: From Barricades to Boardrooms*. New York: Routledge, 2020.

PrivateMilitary.org. "Directory of Private Military Companies." http://www.privatemilitary.org/private_military_companies.html, accessed May 10, 2018.

Regan, Patrick. "Third Party Interventions and the Duration of Intrastate Conflicts." *Journal of Conflict Resolution* 46, no. 1 (2002): 55–73.

Regan, Patrick M., and Aysegul Aydin. "Diplomacy and Other Forms of Intervention in Civil Wars." *Journal of Conflict Resolution* 50, no. 5 (2006): 736–756.

Reuters. "Afghan Government Rejects Proposals to Privatize War." October 5, 2018. https://www.reuters.com/article/us-afghanistan-security/afghan-government-rejects-proposals-to-privatize-war-idUSKCN1MF0IH.

Richards, Paul. "West-African Warscapes: War as Smoke and Mirrors: Sierra Leone, 1991–92, 1994–95, 1995–96." *Anthropological Quarterly* 78, no. 2 (2005): 377–402.

Rodenhauser, Tilman, and Jonathan Cuenoud. "Speaking Law to Business: 10-Year Anniversary of the Montreux Document on PMSCs." Law and Conflict, September 17, 2018. https://blogs.icrc.org/law-and-policy/2018/09/17/speaking-law-business-10-year-anniversary-montreux-document-pmscs/.

Ross, Michael. "A Closer Look at Oil, Diamonds, and Civil War." *Annual Review of Political Science* 9 (2006): 265.

Ross, Michael L. "What Do We Know about Natural Resources and Civil War?" *Journal of Peace Research* 41, no. 3 (2004): 337–356.

Sambanis, Nicholas. "What Is Civil War? Conceptual and Empirical Complexities of an Operational Definition." *Journal of Conflict Resolution* 48, no. 6 (2004): 814–858.

Sarkees, Meredith Reid. "The COW Typology of War: Defining and Categorizing Wars," 32. The Correlates of War Project. https://correlatesofwar.org/data-sets/COW-war/

the-cow-typology-of-war-defining-and-categorizing-wars/view, accessed May 10, 2019.

Sarkees, Meredith Reid, and Frank Wayman, eds. *Resort to War: 1816–2007*. Washington, DC: CQ Press, 2010.

Saxena, Sandeep, and Sami Ylaoutinen. "Managing Budgetary Viraments," 1–20. International Monetary Fund, April 18, 2016. https://www.imf.org/en/Publications/ TNM/Issues/2016/12/31/Managing-Budgetary-Virements-43850.

Scahill, Jeremy. *Blackwater: The Rise of the World's Most Powerful Mercenary Army.* New York: Nation Books, 2007.

Schaub, Gary, Jr., and Volker Franke. "Contractors as Military Professionals?" *Parameters: U.S. Army War College* 39, no. 4 (2009): 88–104.

Schearer, David. "Private Military Force and Challenges for the Future." *Cambridge Review of International Affairs* 13, no. 1 (1999): 80–94.

Schedler, Andreas. "Mexico: Transition to Civil War Democracy." In *Politics in the Developing World*, edited by Peter Burnell, Lise Rakner, and Vicky Randall, 336–346. New York: Oxford University Press, 2014.

Scheimer, Michael. "Separating Private Military Companies from Illegal Mercenaries in International Law: Proposing an International Convention for Legitimate Military and Security Support That Reflects Customary International Law." *American University International Law Review* 24, no. 3 (2009): 609–646.

Schleub, Mark. "Alan Grayson, Candidate and Judge, Gets Sanctioned from the Judge." *Orlando Sentinel*, October 30, 2008. http://articles.orlandosentinel.com/2008–10-30/ news/grayson30_1_grayson-iraq-war-judge.

Schwartz, Moshe, and Jennifer Church. "Department of Defense's Use of Contractors to Support Military Operations: Background, Analysis, and Issues for Congress," 1–32. Congressional Research Service, May 17, 2013. https://fas.org/sgp/crs/natsec/R43074. pdf.

Sherman, Jake. "The Markets for Force in Afghanistan." In *The Market for Force: Privatization of Security across World Regions*, edited by Molly Dunigan and Ulrich Petersohn, 103–118. Philadelphia: University of Pennsylvania Press, 2015.

Singer, Peter. "The Dark Truth about Blackwater." Brookings, October 2, 2007. https:// www.brookings.edu/articles/the-dark-truth-about-blackwater/.

Singer, Peter W. *Corporate Warriors: The Rise of the Privatized Military Industry.* Ithaca, NY: Cornell University Press, 2003.

Singer, Peter W. "Corporate Warriors: The Rise of the Privatized Military Industry and Its Ramifications for International Security." *International Security* 26, no. 3 (2001–2002): 186–220.

Smith, David. "South Africa's Ageing White Mercenaries Who Helped Turn Tide on Boko Haram." *The Guardian*, April 14, 2015. https://www.theguardian.com/world/2015/apr/ 14/south-africas-ageing-white-mercenaries-who-helped-turn-tide-on-boko-haram.

Smith, Eugene. "The New Condottieri and US Policy: The Privatization of Conflict and Its Implications." *Parameters* 32, no. 4 (2002): 104–119.

Smith, Martin. "Private Warriors," Frontline, 2005. http://www.shoppbs.pbs.org/wgbh/ pages/frontline//shows/warriors/etc/script.html, accessed July 18, 2019.

Somers, Mark, and Jose C. Casal. "Type of Wrongdoing and Whistle-Blowing: Further Evidence That Type of Wrongdoing Affects the Whistle-Blowing Process." *Public Personnel Management* 40, no. 2 (2011): 151–163.

Spear, Joanna. *Market Forces: The Political Economy of Private Military Companies.* Oslo: FAFO, 2000. https://www.fafo.no/media/com_netsukii/531.pdf.

Spearin, Christopher. "Since You Left: United Nations Peace Support, Private Military and Security Companies, and Canada." *International Journal: Canada's Journal of Global Policy Analysis* 73, no. 1 (2018): 68–84.

Spencer, Geoff. "Government Suspends Mercenary Contract after Violent Demonstration." Associated Press, March 20, 1997. https://apnews.com/2789a5b9f41 da6faf45b531f607883e8.

Springer, Nathan. "Stabilizing the Debate between Population-Centric and Enemy-Centric Counterinsurgency: Success Demands a Balanced Approach." MA thesis, U.S. Army Command and General Staff College, 2011.

Sturgis, Sue. "Blackwater's Iraq Contract Renewed despite Ongoing Massacre Probe." *Facing South*, April 7, 2008. https://www.facingsouth.org/2008/04/blackwaters-iraq-contract-renewed-despite-ongoing-massacre-probe.html.

Taylor, Andrew. "Trump's Use of Military Money for Border Wall Survives Senate Test." *Military Times*, October 17, 2019. https://www.militarytimes.com/news/pentagon-congress/2019/10/18/trumps-use-of-military-money-for-border-wall-survives-senate-test/.

Themnér, Lotta, and Peter Wallensteen. "Armed Conflict, 1946–2012." *Journal of Peace Research* 50, no. 4 (2013): 509–521.

Thompson, John B. "The New Visibility." *Theory, Culture and Society* 22, no. 6 (2005): 31–51.

Thyne, Clayton. "Third Party Intervention and the Duration of Civil Wars: The Role of Unobserved Factors," 1–39. Working Paper, 2008. https://www.researchgate.net/publication/228429944_Third_Party_Intervention_and_the_Duration_of_Civil_Wars_The_Role_of_Unobserved_Factors, accessed January 12, 2012.

Tir, Jaroslav, and Michael Jasinski. "Domestic-Level Diversionary Theory of War: Targeting Ethnic Minorities." *Journal of Conflict Resolution* 52, no. 5 (2008): 641–664.

Tkach, Benjamin. "Corporate Security and Conflict Outcomes." PhD diss., Texas A&M University, 2013.

Tkach, Benjamin. "Private Military and Security Companies, Contract Structure, Market Competition, and Violence in Iraq." *Conflict Management and Peace Science* 36, no. 3 (2019): 291–311.

United Nations High Commissioner for Refugees. "Diyala Governorate Assessment Report," 1–48. November 2016. https://www.unhcr.org/462cbf672.pdf, accessed July 28, 2019.

United Nations Human Rights Office of the High Commissioner. "International Convention against the Recruitment, Use, Financing and Training of Mercenaries." October 20, 2001. https://www.ohchr.org/EN/ProfessionalInterest/Pages/Mercenaries.aspx.

United Nations Office of Human Resources Management. "Working Together, Putting Ethics into Work," 1–11. November 20, 2008. https://docplayer.net/21479566-Working-together-putting-ethics-to-work-contents.html.

United States Government Accountability Office. "Factors DoD Considers When Choosing Best Value Processes Are Consistent with Guidance for Selected Acquisitions," 1–31. US Government Accountability Office, July 2014. https://www.gao.gov/assets/670/665124.pdf.

United States Government Accountability Office. "Rebuilding Iraq: DOD and State Department Have Improved Oversight and Coordination of Private Security Contractors in Iraq, but Further Actions Are Needed to Sustain Improvements," 1–66. July 2008. https://www.gao.gov/new.items/d08966.pdf.

Valentino, Benjamin, Paul Huth, and Dylan Balch-Lindsay. "Draining the Sea: Mass Killings and Guerrilla Warfare." *International Organization* 58, no. 2 (2004): 375–407.

Vallings, Claire, and Magui Moreno-Torres. "Drivers of Fragility: What Makes States Fragile?," 1–31. Department for International Development UK, April 2005. https://webcache.googleusercontent.com/search?q=cache:H6nEcSvaeX0J:https://ageconsearch.umn.edu/bitstream/12824/1/pr050007.pdf+&cd=15&hl=en&ct=clnk&gl=us.

Varian, Hal. "Monitoring Agents with Other Agents." *Journal of Institutional and Theoretical Economics* 146, no. 1 (1990): 153–174.

"Verbatim," *Time Magazine*, September 27, 2010, 3.

Vinci, Anthony. *Armed Groups and the Balance of Power: The International Relations of Terrorists, Warlords, and Insurgents.* New York: Routledge, 2009.

Visiongain. "Lead Analyst Says: Private Military Security Services Market Set to Grow to $420 Bn by 2029," March 13, 2019. https://www.globenewswire.com/news-release/2019/03/13/1752698/0/en/Lead-analyst-says-Private-military-security-services-market-set-to-grow-to-420-bn-by-2029.html.

Voinea, Cosmina Lelia, and Hans van Kranenburg. "Media Influence and Firms Behavior: A Stakeholder Management Perspective." *International Business Research* 10, no. 10 (2017): 23–38.

von Boemcken, Marc. *Between Security Markets and Protection Rackets: Formations of Political Order.* Opladen, Germany: Budrich UniPress Ltd., 2013.

Wahman, Richard. "Iraq Security Firm Joins Bidding for Wall Street's Favorite Detective Agency." *The Guardian*, March 13, 2010. https://www.theguardian.com/business/2010/mar/14/kroll-control-risks-bidding-war.

Walter, Barbara. "The Critical Barrier to Civil War Settlement." *International Organization* 51, no. 3 (1997): 335–364.

Walter, Barbara. "Bargaining Failures and Civil War." *Annual Review of Political Science* 12, no. 1 (2009): 243–261.

Weintraub, Arlene. "10 Questions to Ask before Taking Your Company Public." *Entrepreneur*, July 25, 2013. http://www.entrepreneur.com/article/227487.

Whyte, Dave. "Lethal Regulation: State Corporate Crime and the United Kingdom Government's New Mercenaries." *Journal of Law and Society* 30, no. 4 (2003): 575–600.

Wildavsky, Aaron, and Naomi Caiden. *The New Politics of the Budgetary Process.* New York: Pearson/Longman, 2004.

Williamson, Oliver. "Public and Private Bureaucracies: A Transaction Cost Economics Perspective." *Journal of Law, Economics, and Organization* 15, no. 1 (1999): 306–342.

Wood, Elisabeth Jean. *Insurgent Collective Action and Civil War in El Salvador.* New York: Cambridge University Press, 2003.

World Bank. "World Development Indicators." http://databank.worldbank.org/ddp/home.do?Step=12&id=4&CNO=2, accessed October 10, 2019.

Zagorin, Adam. "A 'Mutiny' in Kabul: Guards Allege Security Problems Have Put Embassy at Risk." Project on Government Oversight, January 17, 2013. https://www.

pogo.org/investigation/2013/01/mutiny-in-kabul-guards-allege-security-problems-have-put-embassy-at-risk/.

Zenko, Micah. "Mercenaries Are the Silent Majority of Obama's Military." *Foreign Policy*, May 18, 2016. https://foreignpolicy.com/2016/05/18/private-contractors-are-the-silent-majority-of-obamas-military-mercenaries-iraq-afghanistan/.

Zhukov, Yuri. "A Theory of Indiscriminate Violence." PhD diss., Harvard University, 2014.

# Index

Tables and figures are indicated by *t* and *f* following the page number

*For the benefit of digital users, indexed terms that span two pages (e.g., 52–53) may, on occasion, appear on only one of those pages.*